AF706863

Reliability, Quality, and Safety for Engineers

Reliability, Quality, and Safety for Engineers

B.S. Dhillon

CRC Press
Taylor & Francis Group
Boca Raton London New York

CRC Press is an imprint of the
Taylor & Francis Group, an **informa** business

CRC Press
Taylor & Francis Group
6000 Broken Sound Parkway NW, Suite 300
Boca Raton, FL 33487-2742

First issued in paperback 2019

© 2005 by Taylor & Francis Group, LLC
CRC Press is an imprint of Taylor & Francis Group, an Informa business

No claim to original U.S. Government works

ISBN-13: 978-0-8493-3068-1 (hbk)
ISBN-13: 978-0-367-39360-1 (pbk)
Library of Congress Card Number 2004051933

This book contains information obtained from authentic and highly regarded sources. Reasonable efforts have been made to publish reliable data and information, but the author and publisher cannot assume responsibility for the validity of all materials or the consequences of their use. The authors and publishers have attempted to trace the copyright holders of all material reproduced in this publication and apologize to copyright holders if permission to publish in this form has not been obtained. If any copyright material has not been acknowledged please write and let us know so we may rectify in any future reprint.

Except as permitted under U.S. Copyright Law, no part of this book may be reprinted, reproduced, transmitted, or utilized in any form by any electronic, mechanical, or other means, now known or hereafter invented, including photocopying, microfilming, and recording, or in any information storage or retrieval system, without written permission from the publishers.

For permission to photocopy or use material electronically from this work, please access www.copyright.com (http://www.copyright.com/) or contact the Copyright Clearance Center, Inc. (CCC), 222 Rosewood Drive, Danvers, MA 01923, 978-750-8400. CCC is a not-for-profit organization that provides licenses and registration for a variety of users. For organizations that have been granted a photocopy license by the CCC, a separate system of payment has been arranged.

Trademark Notice: Product or corporate names may be trademarks or registered trademarks, and are used only for identification and explanation without intent to infringe.

Library of Congress Cataloging-in-Publication Data

Dhillon, B. S.
Reliability, quality, and safety for engineers / by B.S. Dhillon.
p. cm.
Includes bibliographical references and index.
ISBN 0-8493-3068-8 (alk. paper)
1.Reliability (Engineering) 2. Engineering design. 3. Industrial safety. I. Title.

TS173.D495 2004
620'.00452--dc22 2004051933

Visit the Taylor & Francis Web site at
http://www.taylorandfrancis.com

and the CRC Press Web site at
http://www.crcpress.com

Dedication

This book is affectionately dedicated to Mr. H.E. Coles, M.Sc. (Eng.), of Widnes, England, for inspiring the author to pursue university education.

Preface

Today, billions of dollars are being spent annually to produce new products using modern technologies. Many of these products are highly sophisticated and contain millions of parts. For example, a Boeing 747 Jumbo Jet plane is made up of approximately 4.5 million parts including fasteners. Needless to say, reliability, quality, and safety of systems such as this have become more important than ever before. Global competition and other factors are forcing manufacturers to produce highly reliable, good quality, and safe products.

It means that there is a definite need for the reliability, quality, and safety professionals to work closely during product design and other phases. To achieve this goal, it is essential that they understand to a certain degree each other's discipline. At present, to the author's knowledge, there is one journal but no book that only covers the topics of reliability, quality, and safety within its framework. It means, at present, to gain knowledge of each other's specialties, these specialists must study various books, reports, or articles on each of the topics in question. This approach is time-consuming and rather difficult because of the specialized nature of the material involved.

This book is an attempt to meet the need for a single volume that considers all these three topics. The emphasis of the book is on the structure of concepts rather than on mathematical rigor and minute details. The material covered is treated in such a manner that the reader needs no previous knowledge to understand it. The sources of most of the material presented are given in the reference section at the end of each chapter for the benefit of the reader if he/she wishes to delve deeper into particular topics. At appropriate places, the book contains examples along with their solutions and at the end of each chapter there are numerous problems to test reader comprehension. This will allow the volume to be used as a text.

The book is composed of 15 chapters. Chapter 1 presents the need for reliability, quality, and safety, the historical aspects of reliability, quality, and safety, important terms and definitions, and useful information sources. Chapter 2 reviews mathematical concepts useful for understanding subsequent chapters. It covers topics such as Boolean algebra laws, probability definition and properties, useful mathematical definitions, and statistical distributions. Various introductory aspects of reliability are presented in Chapter 3. Chapter 4 is devoted to reliability evaluation of networks with constant unit reliabilities. In contrast, Chapter 5 presents reliability evaluation of networks with time-dependent unit reliabilities.

A number of useful reliability evaluation methods are presented in Chapter 6. Some of these methods are fault tree analysis, failure modes and effect analysis, decomposition method, Markov method, and network reduction approach. Chapters 7 and 8 present two important topics of reliability: reliability testing and reliability management and costing, respectively.

Chapter 9 presents various introductory aspects of quality including design for quality, quality design characteristics, quality goals, and quality assurance system elements. A total of nine quality analysis methods are presented in Chapter 10. These methods include quality control charts, Pareto diagram, quality function deployment, scatter diagram, and design of experiments. Chapter 11 presents the topics of quality management and costing useful for establishing an effective quality control program.

Chapter 12 presents various introductory aspects of safety including safety-related facts and figures, engineers and safety, product hazard classifications, and product liability. Chapter 13 presents 10 methods for performing various types of safety analysis. Topics of safety management and costing are presented in Chapter 14. Chapter 15 presents various important aspects of robot, software, and medical device safety.

This book will be useful to many people including design engineers, system engineers, manufacturing engineers, reliability specialists, quality professionals, safety engineers, engineering administrators, graduate and senior undergraduate students of engineering, researchers and instructors of reliability, quality, and safety, and engineers-at-large.

The author is deeply indebted to many individuals including colleagues, students, and friends for their invisible inputs and encouragement throughout the project. I thank my children Jasmine and Mark for their patience and intermittent disturbances leading to desirable coffee and other breaks. Last, but not the least, I thank my other half, friend, and wife, Rosy, for typing various portions of this book and other related materials, and for her timely help in proofreading.

B.S. Dhillon
Ottawa, Ontario

Author Biography

Dr. B.S. Dhillon is a professor of Mechanical Engineering at the University of Ottawa. He has served as Chairman/Director of the Mechanical Engineering Department/Engineering Management Program for over 10 years at the same institution. He has published over 310 articles on reliability, engineering management, safety, etc. He is or has been on the editorial boards of seven international scientific journals. In addition, Dr. Dhillon has written 27 books on various aspects of reliability, safety, quality, engineering management, and design published by Wiley (1981), Van Nostrand (1982), Butterworth (1983), Marcel Dekker (1984), Pergamon (1986), etc. Many of his books on reliability have been translated into languages such as German, Russian, and Chinese. He has served as General Chairman of two international conferences on reliability and quality control held in Los Angeles and Paris in 1987.

Professor Dhillon has served as a consultant to various organizations and bodies and has many years of experience in the industrial sector. At the University of Ottawa, he has been teaching engineering management, reliability, quality, safety, and related areas for over 24 years and he has also lectured in over 50 countries, including keynote addresses at various scientific conferences held in North America, Europe, Asia, and Africa. Professor Dhillon attended the University of Wales where he received a B.S. in electrical and electronic engineering and an M.S. in mechanical engineering. He received a Ph.D. in industrial engineering from the University of Windsor.

Contents

1

Introduction

1.1 Need for Reliability, Quality, and Safety

Reliability is increasingly becoming important during the design of engineering systems, as our daily lives and schedules are more dependent than ever before on the satisfactory functioning of these systems. Some examples of these systems are computers, trains, automobiles, aircraft, and space satellites.

Some of the specific factors that are playing a key role in increasing the importance of reliability in designed systems include system complexity and sophistication, competition, increasing number of reliability/safety/quality-related lawsuits, public pressures, high acquisition cost, the past well-publicized system failures, and loss of prestige.

The importance of quality in business and industry is increasing rapidly because of factors such as competition, growing demand from customers for better quality, increasing number of quality-related lawsuits, and the global economy. Nonetheless, the cost of quality control accounts for around 7–10% of the total sales revenue of manufacturers [1]. Today, companies are faced with reducing this amount and at the same time improving the quality of products and services for their survival in the Internet economy.

Today, safety has become a critical issue because each year a very large number of people die and get seriously injured due to workplace and other accidents. For example, in the U.S. according to the National Safety Council (NSC), there were 93,400 deaths and a vast number of disabling injuries due to accidents in 1996 [2]. The total cost of these accidents was estimated to be around $121 billion. Some of the other factors that are also playing an instrumental role in demanding the need for better safety are government regulations, public pressures, and increasing number of lawsuits.

Today, billions of dollars are being spent annually to produce new products using modern technologies. Many of these products are highly sophisticated and contain millions of parts. For example, a Boeing 747 jumbo jet is made up of approximately 4.5 million parts including fasteners. Needless to say, reliability, quality, and safety of systems such as this have become more important than ever before. Global competition and other factors are forcing manufacturers to produce highly reliable, good quality, and safe products.

It means that there is a definite need for the reliability, quality, and safety professionals to work closely during design and other phases. To achieve this goal it is essential that they have to a certain degree an understanding of each other's discipline. Once each of these professionals has the necessary knowledge of each other's discipline, many of their work-related difficulties will disappear or at least be reduced to a tolerable level; thus resulting in better reliability, quality, and safety of end products or systems.

1.2 History

This section presents an overview of historical developments in reliability, quality, and safety, separately.

1.2.1 Reliability

The history of reliability may be traced back to World War II, when the Germans first applied reliability concepts to improve reliability of their V1 and V2 rockets. During the period 1945–1950, the U.S. Department of Defense conducted various studies concerning the failure of electronic equipment, equipment maintenance and repair cost, etc. As the result of these studies, it formed an ad hoc committee on reliability in 1950. In 1952, the committee was transformed to a permanent group called the advisory group on the reliability of electronic equipment (AGREE).

In 1954, a national symposium on reliability and quality control was held for the first time in the U.S. and in 1956, the first commercially available book entitled "Reliability Factors for Ground Electronic Equipment" was published [3]. In 1957, the AGREE report and the first military reliability specification entitled "Reliability Assurance Program for Electronic Equipment" were released [4].

Ever since the 1950s many other developments in reliability have taken place and various researchers and authors have contributed to the field of reliability engineering. A detailed history of reliability engineering is available in Ref. [4].

1.2.2 Quality

The history of quality or quality control may be traced back to the construction of the pyramids of ancient Egypt (1315–1090 B.C.) in which quality-related principles were unquestionably practiced; particularly with respect to workmanship, product size, and materials. In the 12th century AD quality standards were established by the guilds [5].

In the modern times (i.e., by 1907), the Western Electric Company in the U.S. used basic quality control principles in design, manufacturing, and installation. In 1916, C.N. Frazee of Telephone Laboratories applied statistical methods to inspection problems. The term "Quality Control" was coined by Radford [6] in 1917. In 1924, Walter A. Shewhart of Western Electric developed quality control charts.

In 1946, the American Society for Quality Control (ASQC) was formed. Over the years, many researchers, authors, and organizations have contributed to the field of quality. A detailed history of the quality field is given in Refs. [4, 5].

1.2.3 Safety

The history of safety may be traced back to the Code of Hammurabi (2000 B.C.) developed by a Babylonian ruler named Hammurabi. In the modern times, a patent was awarded for first barrier safeguard in the U.S. in 1868 [7]. In 1893, the Railway Safety Act was passed by the U.S. Congress and in 1912, the cooperative Safety Congress met in Milwaukee, Illinois [7, 8].

In 1931, the first commercially available book entitled "Industrial Accident Prevention" was published [9]. In 1947, a paper entitled "Engineering for Safety" was presented to the Institute of Aeronautical Sciences [10]. It emphasized the importance of designing safety into airplanes. In 1962, Exhibit 62–41 entitled "System Safety Engineering for the Development of Air Force Ballistic Missiles" was released by the U.S. Air Force. In 1970, the U.S. Congress passed the occupational safety and health act (OSHA).

Over the years, many organizations, researchers, and authors have contributed to the development of the safety field. The history of safety is discussed in more detail in Ref. [11].

1.3 Terms and Definitions

There are a large number of terms and definitions used in reliability, quality, and safety. This section presents some of the commonly used terms and definitions in these three areas taken from the published literature [12–18].

- **Reliability.** This is the probability that an item will perform its specified mission satisfactorily for the stated time when used according to the specified conditions.
- **Quality.** This is the degree to which an item, function, or process satisfies the user's and customer's requirements.
- **Safety.** This is conservation of human life and its effectiveness, and the prevention of damage to items as per specified mission requirements.

- **Failure.** This is the inability of a product or an item to operate within the specified guidelines.
- **Mission time.** This is the time during which the product/item is carrying out its defined mission.
- **Redundancy.** This is the existence of more than one means to accomplish a specified function.
- **Failure mode.** This is the abnormality of items/parts performance which causes the item to be considered as failed.
- **Hazard rate.** This is the rate of change of the number of items that have failed over the number of items that have survived at a specific time.
- **Process inspection.** This is the intermittent examination and measurement with emphasis on the checking of processing variables.
- **Process average quality.** This is the expected quality of items from a specific process normally estimated from first sample results of previous inspection lots.
- **Sample.** This is a group of items chosen randomly and normally from a lot.
- **Sample size.** This is the number of items chosen randomly from a lot to comprise a single sample.
- **Quality control.** This is a management function, whereby control of the quality of manufactured items and raw materials is exercised to prevent the production of defective items.
- **Control chart.** This is the chart that presents control limits.
- **Accident.** This is an undesired and unplanned act.
- **Safety management.** This is the accomplishment of safety through the effort of other people.
- **Safety process.** This is a series/set of procedures followed to enable an item's safety requirements to be identified and satisfied.
- **Safety assessment.** This is quantitative/qualitative determination of safety.
- **Unsafe condition.** This is any condition (i.e., under the right set of conditions) that will lead to an accident.
- **Confidence limits.** These are the upper and lower extremes of the confidence interval.
- **Quality measure.** This is a quantitative measure of the characteristics and features of an item or service.
- **Quality management.** This is the totality of functions involved in determining and achieving quality.
- **Reliability demonstration.** This is evaluating item capability to satisfy required reliability by actually operating it.

- **Safety function.** This is a function carried out by items which must operate on at least a required minimum level to prevent the occurrence of accidents.
- **Reliability model.** This is a model for predicting, estimating, or assessing reliability.

1.4 Useful Information on Reliability, Quality, and Safety

There are many sources for obtaining reliability, quality, and safety-related information. Some of the most useful sources are presented below under a number of distinct categories.

1.4.1 Journals

- International Journal of Reliability, Quality, and Safety Engineering
- Reliability Engineering and System Safety
- Quality and Reliability Engineering International
- Quality and Reliability Management
- Reliability Review
- IEEE Transactions on Reliability
- Microelectronics Reliability
- Quality Progress
- Technometrics
- International Journal of Health Care Quality Assurance
- Professional Safety
- National Safety News
- Safety Management Journal
- Journal of Safety Research
- Nuclear Safety
- Accident Prevention
- Air Force Safety Journal
- Accident Analysis and Prevention
- Hazard Prevention
- Quality Assurance
- Journal of Quality Technology
- Quality Review
- Industrial Quality Control

- The Quality Circle Journal
- Safety Surveyor

1.4.2 Standards

- MIL-STD-721, Definitions of Terms for Reliability and Maintainability, Department of Defense, Washington, D.C.
- MIL-STD-756, Reliability Modeling and Prediction, Department of Defense, Washington, D.C.
- MIL-STD-785, Reliability Program for Systems and Equipment, Development and Production, Department of Defense, Washington, D.C.
- MIL-STD-2074, Failure Classification for Reliability Testing, Department of Defense, Washington, D.C.
- MIL-STD-1629, Procedures for Performing Failure Mode, Effects and Criticality Analysis, Department of Defense, Washington, D.C.
- MIL-STD-2155, Failure Reporting, Analysis and Corrective Action System (FRACAS), Department of Defense, Washington, D.C.
- MIL-STD-781, Reliability Design, Qualification and Production Acceptance Tests: Exponential Distribution, Department of Defense, Washington, D.C.
- ANSI/ASQC A3, Quality Systems Terminology, American National Standards Institute (ANSI), New York.
- ANSI/ASQC B1, Guide for Quality Control, American National Standards Institute (ANSI), New York.
- ANSI/ASQC B2, Control Chart Method for Analyzing Data, American National Standards Institute, New York.
- ANSI/ASQC E2, Guide to Inspection Planning, American National Standards Institute, New York.
- ANSI/ASQC A1, Definitions, Symbols, Formulas, and Tables for Quality Charts, American National Standards Institute, New York.
- MIL-HDBK-53, Guide for Sampling Inspection, Department of Defense, Washington, D.C.
- MIL-STD-105, Sampling Procedures and Tables for Inspection by Attributes, Department of Defense, Washington, D.C.
- MIL-STD-690, Failure Rate Sampling Plans and Procedures, Department of Defense, Washington, D.C.
- MIL-STD-52779, Software Quality Assurance Program Requirements, Washington, D.C.

- MIL-STD-882, Systems Safety Program for System and Associated Subsystem and Equipment-Requirements, Department of Defense, Washington, D.C.
- MIL-STD-58077, Safety Engineering of Aircraft System, Associated Subsystem and Equipment: General Requirements, Department of Defense, Washington, D.C.
- IEC 60950, Safety of Information Technology Equipment, International Electro-Technical Commission (IEC), Geneva, Switzerland.
- DEF-STD-00-55-1, Requirements for Safety-Related Software in Defense Equipment, Department of Defense, Washington, D.C.
- IEC-601-1: 1988, Medical Electrical Equipment Part I: General Requirements for Safety, International Electro-Technical Commission, Geneva, Switzerland.
- ASTM F1208, Standard Specification for Minimum Performance and Safety Requirements for Anesthesia Breathing Systems, American Society for Testing Materials (ASTM), Philadelphia.
- MIL-STD-790, Reliability Assurance Program for Electronic Parts Specifications, Department of Defense, Washington, D.C.
- MIL-HDBK-338, Electronics Reliability Design Handbook, Department of Defense, Washington, D.C.
- MIL-STD-337, Design to Cost, Department of Defense, Washington, D.C.

1.4.3 Conference Proceedings

- Proceedings of the Annual Reliability and Maintainability Symposium.
- Proceedings of the European Conferences on Safety and Reliability.
- Transactions of the American Society for Quality Control (Conference proceedings).
- Proceedings of the Annual International Reliability, Availability, and Maintainability Conferences for the Electric Power Industry.
- Proceedings of the International Conferences on Probabilistic Safety Assessment and Management.
- Proceedings of the European Organization for Quality Conferences.
- Proceedings of the System Safety Conferences.
- Proceedings of the ASCE Specialty Conferences on Probabilistic Mechanics and Structural Reliability.
- Proceedings of the ISSAT International Conferences on Reliability and Quality in Design.
- Proceedings of the Institute of Quality Assurance Conferences, U.K.

1.4.4 Organizations

- Reliability Society, IEEE, P.O. Box 1331, Piscataway, New Jersey, U.S.A.
- American Society for Quality Control, 310 West Wisconsin Avenue, Milwaukee, Wisconsin, U.S.A.
- American Society of Safety Engineers, 1800 East Oakton St., Des Plaines, Illinois, U.S.A.
- American Society for Testing and Materials, 1916 Race Street, Philadelphia, Pennsylvania, U.S.A.
- System Safety Society, 14252 Culver Drive, Suite A-261, Irvine, California, U.S.A.
- National Safety Council, 444 North Michigan Avenue, Chicago, Illinois, U.S.A.
- World Safety Organization, P.O. Box No. 1, Lalong Laan Building, Pasay City, Metro Manila, The Philippines.
- National Institute for Occupational Safety and Health (NIOSH), 200 Independence Avenue, SW Washington, D.C., U.S.A.
- System Safety Society, P.O. Box 70, Unionville, Virginia, U.S.A.
- European Organization for Quality, 3 rue du Luxembourg, B-1000, Brussels, Belgium.

1.4.5 Books

- Smith, G.M., *Statistical Process Control and Quality Improvement*, Prentice Hall, Upper Saddle River, New Jersey, 2001.
- Vardeman, S., Jobe, J.M., *Statistical Quality Assurance Methods for Engineers*, John Wiley & Sons, New York, 1999.
- Ryan, T.P., *Statistical Methods for Quality Improvement*, John Wiley & Sons, New York, 2000.
- Evans, J.R., Lindsay, W.M., *The Management and Control of Quality*, West Publishing Company, New York, 1989.
- Kolarik, W.J., *Creating Quality: Process Design for Results*, McGraw-Hill, New York, 1999.
- Gryna, F.M., *Quality Planning and Analysis*, McGraw-Hill, New York, 2001.
- Bentley, J.P., *An Introduction to Reliability and Quality Engineering*, John Wiley & Sons, New York, 1993.
- Birolini, A., *Quality and Reliability of Technical Systems: Theory, Practice, and Management*, Springer-Verlag, New York, 1994.
- Cox, S.J., *Reliability, Safety, and Risk Management: An Integrated Approach*, Butterworth-Heinemann, New York, 1991.

- Ballard, C., *Safety*, Blackbirch Press, San Diego, California, 2003.
- Evans, J.W., Evans, J.Y., *Product Integrity and Reliability in Design*, Springer-verlag, New York, 2001.
- Ayyub, B.M., McCuen, R.H., *Probability, Statistics, and Reliability for Engineers and Scientists*, CRC Press, Boca Raton, Florida, 2003.
- Condra, L.W., *Reliability Improvement with Design of Experiments*, Marcel Dekker, New York, 2001.
- Dhillon, B.S., *Design Reliability: Fundamentals and Applications*, CRC Press, Boca Raton, Florida, 1999.
- Shooman, M.L., *Probabilistic Reliability: An Engineering Approach*, McGraw-Hill, New York, 1968.
- Grant Ireson, W., Coombs, C.F., Moss, R.Y., Editors, *Handbook of Reliability Engineering and Management*, McGraw-Hill, New York, 1996.
- Spellman, F.R., Whiting, N.E., *Safety Engineering: Principles and Practice*, Government Institutes, Rockville, Maryland, 1999.
- Hammer, W., Price, D., *Occupational Safety Management and Engineering*, Prentice Hall, Upper Saddle River, New Jersey, 2001.
- Dhillon, B.S., *Engineering Safety: Fundamentals, Techniques, and Applications*, World Scientific Publishing, River Edge, New Jersey, 2003.
- Dhillon, B.S., *Quality Control, Reliability, and Engineering Design*, Marcel Dekker, New York, 1985.
- Goetsch, D.L., *Occupational Safety and Health*, Prentice Hall, Englewood Cliffs, New Jersey, 1996.
- Handley, W., *Industrial Safety Handbook*, McGraw-Hill, London, 1969.
- Hartman, M.G., Ed., *Fundamentals Concepts of Quality Improvement*, ASQ Quality Press, Milwaukee, Wisconsin, 2002.
- Kemp, K.W., *The Efficient Use of Quality Control Data*, Oxford University Press, New York, 2001.

1.5 Problems

1. Discuss the need for reliability, quality, and safety.
2. Discuss the historical developments in quality area.
3. Discuss the history of the reliability field.
4. Define the following terms:
   - Reliability
   - Quality
   - Safety

5. List at least five organizations concerned with reliability or quality.
6. Discuss historical developments in the areas of safety.
7. Define the following terms:
   - Failure
   - Safety process
   - Safety management
   - Sample
8. Compare reliability with quality.
9. What is the difference between, say, product reliability and product safety.
10. List at least five most important standards on reliability and safety.

References

1. Feigenbaum, A.V., *Total Quality Control*, McGraw-Hill, New York, 1983.
2. Accident Facts, Report, National Safety Council, Chicago, Illinois, 1996.
3. Henney, K., Ed., *Reliability Factors for Ground Electronic Equipment*, McGraw-Hill, New York, 1956.
4. Dhillon, B.S., *Reliability and Quality Control: Bibliography on General and Specialized Areas*, Beta Publishers, Gloucester, Ontario, 1992.
5. Hayes, G.E. and Romig, H.G., *Modern Quality Control*, Collier Macmillan, London, 1977.
6. Radford, G.S., Quality control (control of quality), *Industrial Management*, Vol. 54, pp. 100, 1917.
7. Goetsch, D.L., *Occupational Safety and Health*, Prentice Hall, Englewood Cliffs, NJ, 1996.
8. Hammer, W. and Price, D., *Occupational Safety Management and Engineering*, Prentice Hall, Upper Saddle River, NJ, 2001.
9. Heinrich, H.W., *Industrial Accident Prevention*, 3rd ed., McGraw-Hill, New York, 1950.
10. Roland, H.E. and Moriarty, B., *System Safety Engineering and Management*, John Wiley & Sons, New York, 1983.
11. Dhillon, B.S., *Engineering Safety: Fundamentals, Techniques, and Applications*, World Scientific Publishing, River Edge, NJ, 2003.
12. Naresky, J.J., Reliability definitions, *IEEE Transactions on Reliability*, Vol. 19, pp. 198–200, 1970.
13. MIL-STD-721, Definitions of Effectiveness Terms for Reliability, Maintainability, Human Factors, and Safety, Department of Defense, Washington, D.C.
14. Omdahl, T.P., Ed., *Reliability, Availability, and Maintainability (RAM) Dictionary*, ASQC Quality Press, Milwaukee, WI, 1988.
15. Lester, R.H., Enrich, N.C., and Mottley, H.E., *Quality Control for Profit*, Industrial Press, New York, 1977.

16. *Dictionary of Terms Used in the Safety Profession*, 3rd ed., American Society of Safety Engineers, Des Plaines, IL, 1988.
17. Meulen, M.V.D., *Definitions for Hardware and Software Engineers*, Springer-Verlag, London, 2000.
18. Von Alven, W.H., Ed., *Reliability Engineering*, Prentice Hall, Englewood Cliffs, NJ, 1964.

2

Reliability, Quality, and Safety Mathematics

2.1 Introduction

Just like the development in other areas of science and engineering, mathematics has also played a central role in the development of reliability, quality, and safety fields. The history of mathematics may be traced back to the development of our current number symbols often referred to as the "Hindu-Arabic numeral system" [1]. The first evidence of the use of these symbols is found on stone columns eracted by the Scythian Indian Emperor Asoka in 250 B.C. [1].

The earliest reference to the probability concept is made in a gambler's manual written by Girolamo Cardano (1501–1576) [1]. However, Blaise Pascal (1623–1662) and Pierre Fermat (1601–1665) were the first two individuals who solved the problem of dividing the winnings in a game of chance correctly and independently. Pierre Fermat also introduced the idea of "modern differentiation."

Laplace transforms, frequently used to solve differential equations, were developed by Pierre-Simon Laplace (1749–1827). Needless to say, Refs. [1, 2] provide history of mathematics and probability in more detail. This chapter presents useful reliability, quality, and safety-related mathematical concepts.

2.2 Mode, Median, Range, Arithmetic Mean, Root Mean Square, Mean Deviation, and Standard Deviation

A set of reliability, quality, or safety-related data are useful only if it is analyzed in an effective manner. More specifically, there are certain characteristics of the data that help to describe the nature of a given set of data, thus making better associated decisions.

This section presents a number of statistical measures considered useful in reliability, quality, and safety areas [3, 4].

2.2.1 Mode

This is the most commonly occurring value in a set of data. It is to be noted that the mode of a set of numbers (data) may not exist and even if it does exist it may not be unique. The following examples demonstrate these facts:

Example 2.1

The maintenance department of a manufacturing plant is responsible for the satisfactory operation of a number of engineering-systems. The number of monthly failures associated with these systems over a 12-month period were as follows:

15, 7, 9, 10, 15, 20, 25, 18, 15, 5, 4, and 14

Find the mode of the above set of data values.

In the data set the most frequently occurring number is 15. Thus, it is the mode of the data set.

Example 2.2

Assume that in Example 2.1, the number of monthly failures occurring over a 12-month period were as follows:

4, 9, 11, 13, 14, 16, 18, 20, 21, 24, 25, and 26

Find the mode of the above data set. The data set has no mode.

Example 2.3

In Example 2.1, the number of monthly failures occurring over a 12-month period were as follows:

3, 5, 3, 8, 9, 12, 8, 13, 14, 16, 21, and 25

Find the mode of the above data set.

The data set has two modes, i.e., 3 and 8, and is called bimodal. It means that the data set has no unique mode.

2.2.2 Median

The median of a set of data values arranged in an array (i.e., in order of magnitude) is the very middle value or the average of the two middle values.

Example 2.4

Assume that the following set of numbers represent equipment failures occurring over a 9-month period in an organization:

10, 14, 16, 20, 21, 24, 25, 28, and 30

Find the set median. The middle value of the above set of numbers is 21. Thus, it is the set median.

2.2.3 Range

This is a useful measure of variation or dispersion. The range of a set of data values is the difference between the smallest and the largest values in the set.

Example 2.5

Find the range of the data set given in Example 2.4.

By examining Example 2.4 data set, we conclude that the smallest and the largest values are 10 and 30, respectively. Thus, the range of the set is

$$\begin{aligned} R &= \text{Highest value} - \text{Smallest value} \\ &= 30 - 10 \\ &= 20 \end{aligned}$$

where R is the range. It means that the range of the given data set is 20.

2.2.4 Arithmetic Mean

Often, it is simply referred to as mean and is defined by

$$m = \frac{\sum_{i=1}^{k} M_i}{k} \tag{2.1}$$

where

m = mean value
k = total number of data values
M_i = data value i, for $i = 1, 2, 3, \ldots, k$

Example 2.6

Assume that the inspection department of a manufacturing organization inspected eight identical systems and found 10, 14, 7, 15, 4, 2, 5, and 9 defects in each system. Calculate the average number of defects per system (i.e., arithmetic mean).

Substituting the given data values into Equation (2.1), we get

$$\begin{aligned} m &= \frac{10+14+7+15+4+2+5+9}{8} \\ &= 8.25 \end{aligned}$$

Thus, the average number of defects per system is 8.25. More specifically, the arithmetic mean of the data set is 8.25.

2.2.5 Root Mean Square

This is defined by

$$\mathrm{RMS} = \left[\frac{\sum_{i=1}^{k} M_i^2}{k} \right]^{1/2} \tag{2.2}$$

where RMS is the root mean square.

Example 2.7

Find the root mean square of the data set given in Example 2.6. Using the given data in Equation (2.2) yields

$$\mathrm{RMS} = \left[\frac{10^2 + 14^2 + 7^2 + 15^2 + 4^2 + 2^2 + 5^2 + 9^2}{8} \right]^{1/2}$$
$$= 9.33$$

Thus, the root mean square of the given data set is 9.33.

2.2.6 Mean Deviation

This is one of the commonly used measures of dispersion. It indicates the degree to which given data tend to spread about a mean value. Mean deviation is defined by

$$\mathrm{MD} = \frac{\sum_{i=1}^{k} |M_i - m|}{k} \tag{2.3}$$

where

MD = mean deviation
m = mean value of the given data set
k = total number of data points in a given set of data
M_i = data value i; for $i = 1, 2, 3, \ldots, k$
$|M_i - m|$ = absolute value of the deviation of M_i from m

Example 2.8

Find the mean deviation of the data set given in Example 2.6. The calculated mean value from Example 2.6 is m = 8.25 defects/system. Using the above value and the given data in Equation (2.3), we get

$$\begin{aligned}
MD &= \frac{[|10-8.25| + |14-8.25| + |7-8.25| + |15-8.25| + |4-8.25| + |2-8.25| + |5-8.25| + |9-8.25|]}{8} \\
&= \frac{[1.75 + 5.75 + 1.25 + 6.75 + 4.25 + 6.25 + 3.25 + 0.75]}{8} \\
&= 3.75
\end{aligned}$$

Thus, the mean deviation of the given data set is 3.75.

2.2.7 Standard Deviation

This is probably the most widely used measure of dispersion of data in a given data set about the mean. Standard deviation is defined by

$$\sigma = \left[\frac{\sum_{i=1}^{k} (M_i - \mu)^2}{k} \right]^{1/2} \tag{2.4}$$

where σ is the standard deviation and μ is the mean value.

The following three properties of the standard deviation are associated with the normal distribution discussed subsequently in the chapter:

- 68.27% of the all data values are included between $\mu - \sigma$ and $\mu + \sigma$.
- 95.45% of the all data values are included between $\mu - 2\sigma$ and $\mu + 2\sigma$.
- 99.73% of the all data values are included between $\mu - 3\sigma$ and $\mu + 3\sigma$.

Example 2.9

Find the standard deviation of the Example 2.6 data set. The calculated mean value from Example 2.6 is $\mu = m = 8.25$ defects/system. Using the above result and the given data in Equation (2.4) yields

$$\begin{aligned}
\sigma &= \left[\frac{\begin{array}{l}(10-8.25)^2 + (14-8.25)^2 + (7-8.25)^2 + (15-8.25)^2 + (4-8.25)^2 \\ + (2-8.25)^2 + (5-8.25)^2 + (9-8.25)^2\end{array}}{8} \right]^{1/2} \\
&= \left[\frac{3.06 + 33.06 + 1.56 + 45.56 + 18.06 + 39.06 + 10.56 + 0.56}{8} \right]^{1/2} \\
&= 4.75
\end{aligned}$$

Thus, the standard deviation of the Example 2.6 data set is 4.75.

2.3 Boolean Algebra Laws

Boolean algebra is used to a degree in various reliability and safety studies and is named after George Boole (1813–1864), a mathematician. Some of the Boolean algebra laws are as follows [5–6]:

$$X + Y = Y + X \tag{2.5}$$

where X is a set or an event, Y is a set or an event, and "+" denotes the union of sets or events.

$$X \cdot Y = Y \cdot X \tag{2.6}$$

where dot between X and Y or Y and X denotes the intersection of sets or events. It is to be noted that many times the intersection of events is written without using the dot (e.g., XY) but it still conveys the same meaning.

$$X + X = X \tag{2.7}$$

$$YY = Y \tag{2.8}$$

$$Y + YX = Y \tag{2.9}$$

$$X(X + Y) = X \tag{2.10}$$

$$X(Y + Z) = XY + XZ \tag{2.11}$$

where Z is a set or an event.

$$(X + Y)(X + Z) = X + YZ \tag{2.12}$$

$$X + 0 = X \tag{2.13}$$

2.4 Probability Definition and Properties

Probability may simply be defined as likelihood of occurrence of a given event. Mathematically, it can simply be expressed as follows [7]:

$$P = \frac{\text{Total number of ways an event occurs}}{\text{Total number of occurrence possibilities}} \tag{2.14}$$

where P is the probability.

Nonetheless, in many probability-related books, the probability is defined as follows [8]:

$$P(X) = \lim_{k \to \infty} \left[\frac{N}{k} \right] \tag{2.15}$$

where $P(X)$ is the probability of occurrence of event X and N is the total number of times event X occurs in the k repeated experiments.

Some of the most important event-related properties of probability are as follows [8]:

- The probability of occurrence of event, say Y, is

$$0 \le P(Y) \le 1 \tag{2.16}$$

- The probability of occurrence and nonoccurrence of an event, say Y, is always

$$P(Y) + P(\bar{Y}) = 1 \tag{2.17}$$

  where $P(Y)$ is the probability of occurrence of event Y and $P(\bar{Y})$ is the probability of nonoccurrence of event Y.
- The probability of the union of m mutually exclusive events is

$$P(Y_1 + Y_2 + \cdots + Y_m) = \sum_{i=1}^{m} P(Y_i) \tag{2.18}$$

  where $P(Y_i)$ is the probability of occurrence of event Y_i, for $i = 1, 2, 3, \ldots, m$.
- The probability of the union of m independent events is given by

$$P(Y_1 + Y_2 + \cdots + Y_m) = 1 - \prod_{i=1}^{m} (1 - P(Y_i)) \tag{2.19}$$

- The probability of an intersection of m independent events is given by

$$P(Y_1 Y_2 Y_3 \ldots Y_m) = P(Y_1)P(Y_2)P(Y_3)\ldots P(Y_m) \tag{2.20}$$

Example 2.10

Assume that two items are selected at random from a total of 10 items of which four are defective. Calculate the probability that none of the selected items is defective.

The total number of ways to choose two items from the 10 items are given by

$$TNW = \binom{10}{2} = \frac{10!}{2!(10-2)!} = 45$$

The number of nondefective items = 10 – 4 = 6. Thus, the total number of ways to choose two nondefective items are

$$TNWND = \binom{6}{2} = \frac{6!}{2!(6-2)!} = 15$$

Substituting the above calculated values into Equation (2.14) yields

$$P = \frac{15}{45} = 0.3333$$

Thus, the probability that none of the selected items is defective is 0.3333.

2.5 Useful Mathematical Definitions

This section presents a number of mathematical definitions considered useful to perform various types of reliability, quality, and safety studies.

2.5.1 Probability Density and Cumulative Distribution Functions

For a continuous random variable, the probability density function is defined by

$$f(t) = \frac{dF(t)}{dt} \tag{2.21}$$

where
t = time (i.e., a continuous random variable)
$f(t)$ = probability density function (in reliability work, it is often referred to as failure density function)
$F(t)$ = cumulative distribution function and is expressed by

$$F(t) = \int_{-\infty}^{t} f(x)dx \tag{2.22}$$

and

$$F(\infty) = 1 \tag{2.23}$$

Usually, in reliability work, Equation (2.22) is simply written as

$$F(t) = \int_0^t f(x)\,dx \tag{2.24}$$

2.5.2 Expected Value

The expected value, $E(t)$, of a continuous random variable is defined by

$$E(t) = m = \int_{-\infty}^{\infty} t\,f(t)\,dt \tag{2.25}$$

where $E(t)$ is the expected value of the continuous random variable t and m is the mean value.

Similarly, the expected value $E(t)$ of a discrete random variable t is expressed by

$$E(t) = \sum_{j=1}^{K} t_j f(t_j) \tag{2.26}$$

where K is the total number of discrete values of the random variable t.

2.5.3 Variance

The variance σ^2 (t) of a random variable t is expressed by

$$\sigma^2(t) = E(t^2) - [E(t)]^2 \tag{2.27}$$

or

$$\sigma^2(t) = \int_0^{\infty} t^2 f(t)\,dt - m^2 \tag{2.28}$$

where m is the mean value. In reliability work, it is commonly known as mean time to failure.

TABLE 2.1

Laplace transforms of some functions

No.	$f(t)$	$f(s)$
1	K (a constant)	$\frac{K}{s}$
2	t^a, for a = 0, 1, 2,...	$\frac{a!}{s^{a+1}}$
3	$\alpha_1 f_1 + \alpha_2 f_2(t)$	$\alpha_1 f_1(s) + \alpha_2 f_2(s)$
4	$e^{-\lambda t}$	$\frac{1}{(s+\lambda)}$
5	$te^{-\lambda t}$	$\frac{1}{(s+\lambda)^2}$
6	$\frac{df(t)}{dt}$	$sf(s) - f(0)$

2.5.4 Laplace Transform

This is defined by

$$f(s) = \int_0^\infty f(t)e^{-st}dt \tag{2.29}$$

where s is the Laplace transform variable and $f(s)$ is the Laplace transform of $f(t)$.

Laplace transforms of some frequently occurring functions in reliability, quality, or safety areas are presented in Table 2.1 [9–10].

2.5.5 Laplace Transform: Final-Value Theorem

If the following limits exist, then the final value theorem may be stated as

$$\lim_{t\to\infty} f(t) = \lim_{s\to 0} sf(s) \tag{2.30}$$

Example 2.11

Prove using the following equation that the left-hand side of Equation (2.30) is equal to its right-hand side:

$$f(t) = \frac{\mu}{(\lambda+\mu)} + \frac{\lambda}{(\lambda+\mu)}e^{-(\lambda+\mu)t} \tag{2.31}$$

where λ and μ are constants.

Inserting Equation (2.31) into the left-hand side of Equation (2.30) yields

$$\lim_{t\to\infty}\left[\frac{\mu}{(\lambda+\mu)}+\frac{\lambda}{(\lambda+\mu)}e^{-(\lambda+\mu)t}\right]=\frac{\mu}{(\lambda+\mu)} \tag{2.32}$$

Using Table 2.1, we get the following Laplace transforms of Equation (2.31):

$$f(s)=\frac{\mu}{s(\lambda+\mu)}+\frac{\lambda}{(\lambda+\mu)}\cdot\frac{1}{(s+\lambda+\mu)} \tag{2.33}$$

Substituting Equation (2.33) into the right-hand side of Equation (2.30) yields:

$$\lim_{s\to 0}\left[\frac{s\mu}{s(\lambda+\mu)}+\frac{s\lambda}{(\lambda+\mu)(s+\lambda+\mu)}\right]=\frac{\mu}{\lambda+\mu} \tag{2.34}$$

The right-hand sides of Equation (2.32) and Equation (2.34) are the same. Thus, it proves that the left-hand side of Equation (2.30) is equal to its right-hand side.

2.6 Solving First Order Differential Equations with Laplace Transforms

Solutions to linear first order differential equations are often found by using Laplace transforms, particularly when a set of linear first order differential equations are involved. The following example demonstrates the finding of solutions to a set of differential equations describing an engineering system:

Example 2.12

An engineering system can be in either of the three distinct states: operating normally, failed in open mode, or failed in short mode. The following three differential equations describe the system:

$$\frac{dP_0(t)}{dt}+(\lambda_0+\lambda_s)P_0(t)=0 \tag{2.35}$$

$$\frac{dP_1(t)}{dt}-\lambda_0 P_0(t)=0 \tag{2.36}$$

$$\frac{dP_2(t)}{dt}-\lambda_s P_0(t)=0 \tag{2.37}$$

where

$P_i(t)$ = probability that the system is in state i at time t, for $i = 0$ (operating normally), $i = 1$ (failed in open mode), and $i = 2$ (failed in short mode).

λ_0 = system constant open mode failure rate

λ_s = system constant short mode failure rate

At time $t = 0$, $P_0(0) = 1$, $P_1(0) = 0$, and $P_2(0) = 0$.

Solve differential Equations (2.35–2.37) by using Laplace transforms. Using Table 2.1, Equations (2.35–2.37), and the given initial conditions, we write

$$sP_0(s) - 1 + (\lambda_0 + \lambda_s)P_0(s) = 0 \tag{2.38}$$

$$sP_1(s) - \lambda_0 P_0(s) = 0 \tag{2.39}$$

$$sP_2(s) - \lambda_s P_0(s) = 0 \tag{2.40}$$

Solving Equations (2.38–2.40), we get

$$P_0(s) = \frac{1}{(s + \lambda_0 + \lambda_s)} \tag{2.41}$$

$$P_1(s) = \frac{\lambda_0}{s(s + \lambda_0 + \lambda_s)} \tag{2.42}$$

$$P_2(s) = \frac{\lambda_s}{s(s + \lambda_0 + \lambda_s)} \tag{2.43}$$

The inverse Laplace transforms of Equations (2.41–2.43) are

$$P_0(t) = e^{-(\lambda_0 + \lambda_s)t} \tag{2.44}$$

$$P_1(t) = \frac{\lambda_0}{\lambda_0 + \lambda_s}[1 - e^{-(\lambda_0 + \lambda_s)t}] \tag{2.45}$$

and

$$P_2(t) = \frac{\lambda_s}{\lambda_0 + \lambda_s}[1 - e^{-(\lambda_0 + \lambda_s)t}] \tag{2.46}$$

Equations (2.44–2.46) are the solutions to Equations (2.35–2.37).

2.7 Statistical Distributions

This section presents a number of statistical or probability distributions useful for performing various types of studies in reliability, quality, and safety areas [11].

2.7.1 Normal Distribution

This is one of the most widely used continuous random variable distributions and time to time is referred to as Gaussian distribution, after a German mathematician named Carl Friedrich Gauss (1777–1855). But, the distribution was actually discovered by De Moivre in 1733 [11].

The probability density function of the distribution is expressed by

$$f(t) = \frac{1}{\sigma\sqrt{2\pi}} \exp\left[-\frac{(t-\mu)}{2\sigma^2}\right], \quad -\infty < t < +\infty \tag{2.47}$$

where σ and μ are the distribution parameters (i.e., standard deviation and mean, respectively).

Using Equation (2.47) in Equation (2.22) yields

$$F(t) = \frac{1}{\sigma\sqrt{2\pi}} \int_{-\infty}^{t} \exp\left[-\frac{(x-\mu)^2}{2\sigma^2}\right] dx \tag{2.48}$$

where $F(t)$ is the cumulative distribution function.

Substituting Equation (2.47) into Equation (2.25) yields the following expression for the distribution mean value:

$$\begin{aligned} E(t) = m &= \frac{1}{\sigma\sqrt{2\pi}} \int_{-\infty}^{\infty} t \exp\left[-\frac{(t-\mu)}{2\sigma^2}\right] dt \\ &= \mu \end{aligned} \tag{2.49}$$

2.7.2 Binomial Distribution

This discrete random variable distribution is used in situations where one is concerned with the probabilities of outcome such as the number of occurrences (e.g., failures) in a sequence of given number of trials. More specifically, each trial has two possible outcomes (e.g., success or failure), but the probability of each trial remains unchanged or constant.

The distribution is also known as the Bernoulli distribution after Jakob Bernoulli (1654–1705) [1]. The probability density function of the distribution is expressed by

$$f(y) = \frac{n!}{y!(n-y)!} p^y q^{n-y}, \quad \text{for } y = 0, 1, 2, 3, \ldots, n \tag{2.50}$$

where

y = number of nonoccurrences (e.g., failures) in a total of n trials
p = single trial probability of occurrence (e.g., success)
q = single trial probability of nonoccurrence (e.g., failure)

The cumulative distribution function is given by

$$F(y) = \sum_{i=0}^{y} \frac{n!}{i!(n-i)!} p^i q^{n-i} \tag{2.51}$$

where $F(y)$ is the probability of y or less nonoccurrences (e.g., failures) in n trials.

Using Equation (2.26) and Equation (2.50), the expected value or the mean of the distribution is

$$E(y) = np \tag{2.52}$$

2.7.3 Poisson Distribution

This is another discrete random variable distribution and is used in situations where one is interested in the occurrence of a number of events of the same type. More specifically, the distribution is used when the number of possible events is large but the probability of occurrence over a given time interval is small. Two typical examples of such a situation are the waiting lines and the occurrence of defects.

The distribution is named after Simeon Poisson (1781–1840), a French mathematician and its probability density function is expressed by

$$f(y) = \frac{\lambda^y e^{-\lambda}}{y!}, \quad \text{for } y = 0, 1, 2, \ldots \tag{2.53}$$

where λ is the distribution parameter.

The cumulative distribution function is

$$F(Y) = \sum_{i=0}^{y} \frac{\lambda^i e^{-\lambda}}{i!} \tag{2.54}$$

Using Equation (2.26) and Equation (2.53), the expected value or the mean of the distribution is

$$E(y) = \lambda \tag{2.55}$$

2.7.4 Exponential Distribution

This is one of the simplest continuous random variable distributions widely used in industry, particularly in reliability studies [12]. The probability density function of the distribution is expressed by

$$f(t) = \lambda e^{-\lambda t}, \quad t \geq 0, \lambda > 0 \tag{2.56}$$

where t is time and λ is the distribution parameter.

Substituting Equation (2.56) into Equation (2.24) yields the following expression for the cumulative distribution function:

$$F(t) = 1 - e^{-\lambda t} \tag{2.57}$$

Using Equation (2.56) and Equation (2.25), we get the following expression for the distribution mean value:

$$E(t) = m = \frac{1}{\lambda} \tag{2.58}$$

2.7.5 Rayleigh Distribution

This is another continuous random variable distribution and is named after John Rayleigh (1842–1919) [1]. The distribution is often used in the theory of sound and in reliability studies. Its probability density function is expressed by

$$f(t) = \frac{2}{\alpha^2} t e^{-(t/\alpha)^2}, \quad t \geq 0, \alpha > 0 \tag{2.59}$$

where α is the distribution parameter. Using Equation (2.24) and Equation (2.59), we get the following cumulative distribution function:

$$F(t) = 1 - e^{-(t/\alpha)^2} \tag{2.60}$$

Inserting Equation (2.59) into Equation (2.25) yields the following equation for the distribution mean value:

$$E(t) = m = \alpha\, \Gamma(1.5) \tag{2.61}$$

where $\Gamma(\cdot)$ is the gamma function and is expressed by

$$\Gamma(y) = \int_0^\infty t^{y-1} e^{-t}\, dt, \quad \text{for } y > 0 \tag{2.62}$$

2.7.6 Weibull Distribution

This continuous random variable distribution is named after W. Weibull, a Swedish mechanical engineering professor [13]. The distribution can be used to represent many different physical phenomena. The probability density function of the distribution is defined by

$$f(t) = \frac{\theta\, t^{\theta-1}}{\beta^\theta} e^{-(t/\beta)^\theta}, \quad t \geq 0,\ \theta > 0,\ \beta > 0 \tag{2.63}$$

where θ and β are distribution shape and scale parameters, respectively.

Inserting Equation (2.63) into Equation (2.24) we get the following equation for cumulative distribution function:

$$F(t) = 1 - e^{-(t/\beta)^\theta} \tag{2.64}$$

Exponential and Rayleigh distributions are the special cases of this distribution for $\theta = 1$ and 2, respectively.

Using Equation (2.25) and Equation (2.63), we get the following expression for the Weibull distribution mean value:

$$E(t) = m = \beta\Gamma\left(1 + \frac{1}{\theta}\right) \tag{2.65}$$

2.8 Problems

1. Write an essay on the early developments in mathematics.
2. Define the following two items:
   - Mode
   - Arithmetic mean
3. Find the root mean square of the following set of numbers: 4, 5, 8, 11, 12, and 14.
4. What is the difference between mean deviation and standard deviation?
5. Prove expression (2.12).

6. Define the following:
    - Probability
    - Probability density function
    - Cumulative distribution function
7. What are the special case distributions of the Weibull distribution?
8. Find the Laplace transform of the following function:

$$f(t) = te^{-\lambda t} \tag{2.66}$$

    where t is time and λ is a constant.
9. Prove that the mean value of the Weibull distribution is given by Equation (2.65).
10. Discuss the following distributions:
    - Poisson distribution
    - Binomial distribution

References

1. Eves, H., *An Introduction to the History of Mathematics*, Holt, Rinehart, & Winston, New York, 1976.
2. Owen, D.B., Ed., *On the History of Statistics and Probability*, Marcel Dekker, New York, 1976.
3. Spiegel, M.R., *Statistics*, McGraw-Hill, New York, 1961.
4. Spiegel, M.R., *Probability and Statistics*, McGraw-Hill, New York, 1975.
5. Fault Tree Handbook, Report No. NUREG-0492, U.S. Nuclear Regulatory Commission, Washington, D.C., 1981.
6. Lipschutz, S., *Set Theory*, McGraw-Hill, New York, 1964.
7. Omdahl, T.P., Ed., *Reliability, Availability, and Maintainability (RAM) Dictionary*, ASQC Quality Press, Milwaukee, WI, 1988.
8. Mann, N.R., Schafer, R.E., and Singpurwalla, N.D., *Methods for Statistical Analysis of Reliability and Life Data*, John Wiley & Sons, New York, 1974.
9. Spiegel, M.R., *Laplace Transforms*, McGraw-Hill, New York, 1965.
10. Oberhettinger, F., and Badic, L., *Tables of Laplace Transforms*, Springer-Verlag, New York, 1973.
11. Patel, J.K., Kapadia, C.H., and Owen, D.B., *Handbook of Statistical Distributions*, Marcel Dekker, New York, 1976.
12. Davis, D.J., An analysis of some failure data, *J. Amer. Stat. Assoc.*, June 1952, pp. 113–150.
13. Weibull, W., A statistical distribution function of wide applicability, *J. Appl. Mech.*, Vol. 18, March 1951, pp. 293–297.

3

Introduction to Reliability

3.1 Need for Reliability

Reliability is becoming an important factor during the design of engineering systems/products as our daily lives and schedules are more dependent than ever before on the satisfactory functioning of these systems. Some examples of these systems are trains, computers, aircraft, space satellites, and automobiles.

Usually, the required reliability of items such as these is stated in the design specification. In turn, every effort is made during the design phase to meet this requirement satisfactorily.

Some of the factors that are playing an instrumental role in increasing the importance of reliability in designed systems include competition, public pressures, increasing number of reliability/quality-related lawsuits, complex and sophisticated products, loss of prestige, high acquisition cost, and the past well-publicized system failures. Three of these factors are discussed in more detail below.

- **Complex and sophisticated products.** Over the years engineering products have become more sophisticated and complex. For example, in 1935 a farm tractor had 1200 critical parts and in 1990 the number increased to around 2900. Furthermore, today a typical Boeing 747 jumbo jet airplane is composed of around 4.5 million parts, including fasteners [1].
- **High acquisition cost.** Many engineering products cost millions of dollars (e.g., commercial airplanes, defense systems, and space satellites). Failure of such items could result in loss of millions of dollars.
- **The past well-publicized system failures.** Three examples of these failures are Space Shuttle Challenger disaster, Chernobyl Nuclear Reactor explosion, and Point Pleasant Bridge Disaster [2–4]. These disasters occurred in January 1986, April 1986, and December 1967, respectively.

3.2 Bathtub Hazard Rate Curve

Bathtub hazard rate curve is a well known concept to represent failure behaviour of various engineering items/products because the failure rate of these items changes with time. Its name stem from its shape resembling a bathtub as shown in Figure 3.1. Three distinct regions of the curve are identified in the figure: burn-in region, useful life region, and wear-out region. These regions denote three phases that a newly manufactured product passes through during its life span.

During the burn-in region/period, the product hazard rate (i.e., time dependent failure rate) decreases and some of the reasons for the occurrence of failures during this period are poor workmanship, substandard parts and materials, poor quality control, poor manufacturing methods, incorrect installation and start-up human error, inadequate debugging, incorrect packaging, inadequate processes, and poor handling methods [5]. Other names used for the "burn-in region" are "debugging region," "infant mortality region," and "break-in region."

During the useful life region, the product hazard rate remains constant and the failures occur randomly or unpredictably. Some of the reasons for their occurrence are undetectable defects, abuse, low safety factors, higher random stress than expected, unavoidable conditions, and human errors [5].

During the wear-out region, the product hazard rate increases and some of the reasons for the occurrence of "wear-out region" failures are as follows [5]:

- Poor maintenance
- Wear due to friction
- Wear due to aging
- Corrosion and creep
- Wrong overhaul practices
- Short designed-in life of the product

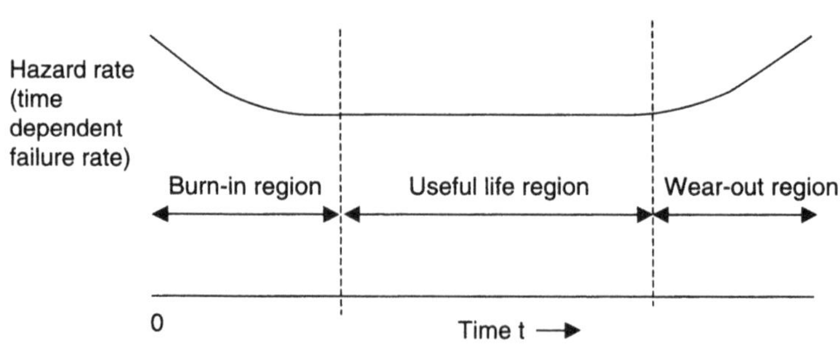

FIGURE 3.1
Bathtub hazard rate curve.

3.3 General and Specific Hazard Rate Functions

The general hazard rate of an item is defined by [6]

$$\begin{aligned}\lambda(t) &= \frac{f(t)}{R(t)} \\ &= \frac{f(t)}{1-F(t)} \\ &= \frac{f(t)}{1-\int_0^t f(t)dt}\end{aligned} \tag{3.1}$$

where

$\lambda(t)$ = item hazard rate (i.e., time t dependent failure rate)
$f(t)$ = item failure density function (probability density function)
$F(t)$ = cumulative distribution function (i.e., the item failure probability at time t)
$R(t)$ = item reliability at time t

The denominator of Equation (3.1) can be expressed as follows:

$$R(t) = 1 - \int_0^t f(t)dt \tag{3.2}$$

Differentiating Equation (3.2) with respect to t, we get

$$\frac{dR(t)}{dt} = -f(t) \tag{3.3}$$

Substituting Equation (3.3) into Equation (3.1) yields

$$\lambda(t) = -\frac{1}{R(t)} \cdot \frac{dR(t)}{dt} \tag{3.4}$$

This expression is quite useful to obtain item hazard rate when the item's reliability function is known.

Hazard rate functions for three specific probability distributions are presented below.

3.3.1 Hazard Rate for Exponential Distribution

The failure density function of the exponential distribution is defined by [6]

$$f(t) = \lambda e^{-\lambda t}, \quad t \geq 0, \lambda > 0 \tag{3.5}$$

where λ is the distribution parameter (in reliability studies, it is known as failure rate) and t is time.

Inserting Equation (3.5) into Equation (3.1), we get the following expression for exponential distribution's hazard rate:

$$\begin{aligned}\lambda(t) &= \frac{\lambda e^{-\lambda t}}{1-\int_0^t \lambda e^{-\lambda t}dt} \\ &= \lambda\end{aligned} \tag{3.6}$$

The right-hand side of Equation (3.6) is independent of time t. Thus, λ is called constant failure rate. It simply means that when an item's times to failure are exponentially distributed, its failure rate is automatically constant.

3.3.2 Hazard Rate for Weibull Distribution

The failure density function of the Weibull distribution is defined by [6]

$$f(t) = \frac{\theta t^{\theta-1}}{\alpha^\theta} e^{-(t/a)^\theta}, \quad t \geq 0, \alpha > 0, \theta > 0 \tag{3.7}$$

where α and θ are the scale and shape parameters, respectively.

Using Equation (3.7) in Equation (3.1) yields the following expression for Weibull distribution's hazard rate:

$$\begin{aligned}\lambda(t) &= \left[\frac{\theta t^{\theta-1}}{\alpha^\theta} e^{-(t/a)^\theta}\right] \Big/ \left[1-\int_0^t \frac{\theta t^{\theta-1}}{\alpha^\theta} e^{-(t/a)^\theta} dt\right] \\ &= \frac{\theta}{\alpha^\theta} t^{\theta-1}\end{aligned} \tag{3.8}$$

For $\theta = 1$ and $\theta = 2$, Equation (3.8) gives hazard rates for exponential and Rayleigh distributions, respectively. For example, Equation (3.8) is identical to Equation (3.6) when $\theta = 1$ and $\alpha = 1/\lambda$. Hazard or failure rate increases linearly with time when times to failures are Rayleigh distributed. This can easily be seen from Equation (3.8) for $\theta = 2$.

3.3.3 Hazard Rate for General Distribution

The failure density function of the general distribution is defined by [7]

$$\begin{aligned}&f(t) = [c\lambda\gamma t^{\gamma-1} + (1-c)\theta t^{\theta-1}\mu e^{\mu t^\theta}][\exp[-c\lambda t^\gamma - (1-c)(e^{\mu t^\theta}-1)]] \\ &\text{for } 0 \leq c \leq 1 \quad \text{and} \quad \gamma, \theta, \mu, \lambda > 0\end{aligned} \tag{3.9}$$

where

t = time,
μ and λ = scale parameters
θ and γ = shape parameters

Substituting Equation (3.9) into Equation (3.1) yields the following expression for general distribution's hazard rate:

$$\lambda(t) = c\lambda\gamma t^{\gamma-1} + (1-c)\,\theta t^{\theta-1}\mu e^{\mu t^{\theta}} \tag{3.10}$$

For $\theta = 1$ and $\gamma = 0.5$, Equation (3.10) gives the bathtub hazard rate curve.

Hazard rates for the following probability distributions are the special cases of Equation (3.10):

- Exponential (i.e., for $c = 1$, $\gamma = 1$)
- Rayleigh (i.e., for $c = 1$, $\gamma = 2$)
- Weibull (i.e., for $c = 1$)
- Extreme value (i.e., for $c = 0$, $\theta = 1$)
- Makeham (i.e., for $\gamma = 1$, $\theta = 1$)

3.4 General and Specific Reliability Functions

The general reliability function can be obtained by using Equation (3.4). Thus, rearranging Equation (3.4), we get

$$-\lambda(t)dt = \frac{1}{R(t)}dR(t) \tag{3.11}$$

Integrating both sides of Equation (3.11) over the time interval [0, t], we get

$$-\int_0^t \lambda(t)dt = \int_1^{R(t)} \frac{1}{R(t)}dR(t) \tag{3.12}$$

since at $t = 0$, $R(t) = 1$.

Evaluating the right-hand side of Equation (3.12) and rearranging the resulting expression yields

$$\ln R(t) = -\int_0^t \lambda(t)dt \tag{3.13}$$

Thus, using Equation (3.13), we get

$$R(t) = e^{-\int_0^t \lambda(t)dt} \tag{3.14}$$

Equation (3.14) is the general expression for the reliability function. It can be used to obtain reliability function of an item when the item's hazard rate is defined by any probability distribution. This is demonstrated below for three specific distributions: exponential, Weibull, and general.

3.4.1 Reliability Function for Exponential Distribution

Substituting Equation (3.6) into Equation (3.14), we get the following expression for exponential distribution's reliability function:

$$\begin{aligned} R(t) &= e^{-\int_0^t \lambda dt} \\ &= e^{-\lambda t} \end{aligned} \tag{3.15}$$

3.4.2 Reliability Function for Weibull Distribution

Using Equation (3.8) in Equation (3.14) yields

$$\begin{aligned} R(t) &= e^{-\int_0^t \frac{\theta}{\alpha^\theta} t^{\theta-1} dt} \\ &= e^{-\left(\frac{t}{\alpha}\right)^\theta} \end{aligned} \tag{3.16}$$

where $R(t)$ is the reliability function for Weibull distribution.

3.4.3 Reliability Function for General Distribution

Inserting Equation (3.10) into Equation (3.14), we get the following expression for general distribution's reliability function:

$$\begin{aligned} R(t) &= e^{-\int_0^t \left[c\lambda\gamma t^{\gamma-1} + (1-c)\theta t^{\theta-1}\mu e^{\mu t^\theta}\right] dt} \\ R(t) &= \exp\left[-c\lambda t^\gamma - (1-c)\left(e^{\mu t^\theta} - 1\right)\right] \end{aligned} \tag{3.17}$$

3.5 Mean Time to Failure

This is an important reliability parameter and it can be obtained by using any of the following three formulas [4]:

$$\text{MTTF} = \int_0^\infty t f(t) dt \tag{3.18}$$

or

$$\text{MTTF} = \int_0^\infty R(t)\,dt \tag{3.19}$$

or

$$\text{MTTF} = \lim_{s \to 0} R(s) \tag{3.20}$$

where
MTTF = mean time to failure
s = Laplace transform variable
$R(s)$ = Laplace transform of the $R(t)$ (i.e., the reliability function)

3.5.1 Equation (3.19) Derivation

Substituting Equation (3.3) into Equation (3.18) yields

$$\text{MTTF} = -\int_0^\infty t \frac{dR(t)}{dt}\,dt \tag{3.21}$$

Integrating Equation (3.21) by parts results in:

$$\text{MTTF} = \int_0^\infty R(t)\,dt - [tR(t)]_0^\infty \tag{3.22}$$

In order to prove that the second right-hand term of Equation (3.22) is equal to zero, we rewrite that term to the following form:

$$\left[\frac{t}{R^{-1}(t)}\right]_0^\infty \tag{3.23}$$

Substituting the limits in the above relationship leads to at $t = 0$, $t\,R(t) = 0$ and at $t = \infty$ for a finite mean, we assume

$$\lim_{t \to \infty}\left[\frac{t}{R^{-1}(t)}\right] = 0 \tag{3.24}$$

Thus, the right-hand term of Equation (3.22) is proven to be equal to zero. Consequently, Equation (3.22) becomes

$$\text{MTTF} = \int_0^\infty R(t)\,dt \tag{3.25}$$

The above equation is identical to Equation (3.19).

3.5.2 Equation (3.20) Derivation

We rewrite Equation (3.19) to the following form [6]

$$\text{MTTF} = \lim_{t \to \infty} \left[\int_0^t R\,(x)\,dx \right] \tag{3.26}$$

The term given in brackets in Equation (3.26) is $f(t)$ and its Laplace transform is

$$\mathcal{L}[f(t)] = \mathcal{L}\left[\int_0^t R(x)dx \right] = \frac{R(s)}{s} \tag{3.27}$$

The Laplace transform of the final value theorem is given by [8]

$$\lim_{t \to \infty} f(t) = \lim_{s \to 0} s\,f(s) \tag{3.28}$$

where $f(s)$ is the Laplace transform of $f(t)$.

Using Equation (3.27) and Equation (3.28) in Equation (3.26) yields

$$\text{MTTF} = \lim_{s \to 0} R(s) \tag{3.29}$$

The above equation is identical to Equation (3.20).

Example 3.1

Using Equation (3.5) and Equation (3.15) prove that Equations (3.18–3.20) give the same result for MTTF.

Substituting Equation (3.5) into Equation (3.18) yields

$$\begin{aligned} \text{MTTF} &= \int_0^\infty t\lambda e^{-\lambda t} dt \\ &= [-te^{-\lambda t}]_0^\infty - \left[-\frac{e^{-\lambda t}}{\lambda}\right]_0^\infty \\ &= \frac{1}{\lambda} \end{aligned} \tag{3.30}$$

Using Equation (3.15) in Equation (3.19) yields

$$\begin{aligned} \text{MTTF} &= \int_0^\infty e^{-\lambda t} dt \\ &= \left[\frac{e^{-\lambda t}}{-\lambda}\right]_0^\infty \\ &= \frac{1}{\lambda} \end{aligned} \tag{3.31}$$

Taking the Laplace transform of Equation (3.15) and then inserting it into Equation (3.20), we get

$$\begin{aligned} \text{MTTF} &= \lim_{s \to 0} \frac{1}{(s+\lambda)} \\ &= \frac{1}{\lambda} \end{aligned} \tag{3.32}$$

Equations (3.30–3.32) are identical; thus it proves that Equations (3.18–3.20) give the same result.

Example 3.2

Assume that times to failure of an item follow Weibull distribution and its hazard rate is defined by

$$\lambda(t) = \frac{mt^{m-1}}{\alpha} \tag{3.33}$$

where

t = time
α = scale parameter
m = shape parameter

Obtain an expression for the item's mean time to failure.
Inserting Equation (3.33) into Equation (3.14) we get

$$R(t) = e^{-\int_0^t \frac{m}{\alpha} t^{m-1} dt} = e^{-\frac{t^m}{\alpha}} \tag{3.34}$$

Using Equation (3.34) in Equation (3.19) yields

$$\text{MTTF} = \int_0^\infty e^{-\frac{t^m}{\alpha}} dt = \frac{\Gamma\left(\frac{1}{m}\right)}{m\left(\frac{1}{\alpha}\right)^{1/m}} \tag{3.35}$$

where $\Gamma(\cdot)$ is the gamma function.
The item's mean time to failure is given by Equation (3.35).

Example 3.3

Assume that a railway engine's constant failure rate λ is 0.0002 failures per hour. Calculate the engine's mean time to failure.
Substituting the given data value into Equation (3.32), we get

$$\text{MTTF} = \frac{1}{0.0002} = 5000\,\text{h}$$

Thus, the railway engine's expected time to failure is 5000 h.

3.6 Failure Rate Estimation

Usually, during the design process, the failure rates of components used in the product are estimated. The overall product/equipment failure rate is calculated by adding all the component failure rates. Usually, the failure rate of electronic parts is calculated by using the MIL-HDBK-217 [9]. This handbook uses an equation of following form to estimate failure rates of various electronic components [9]:

$$\lambda_C = \lambda_b f_q f_e \cdots \tag{3.36}$$

where

λ_C = component failure rate
f_q = factor that accounts for component quality level
f_e = factor that accounts for the influence of the use environment
λ_b = base failure rate; this is normally expressed by a model relating the temperature and electrical stresses influence on the component under consideration

The base failure rate is obtained by using the following equation, for many electronic parts:

$$\lambda_b = C\exp\left[-\frac{e_a}{kT_a}\right] \tag{3.37}$$

where

k = Boltzmann's constant
T_a = absolute temperature
e_a = activation energy for the process
C = constant

Usually, during bid proposal and early design phases the electronic equipment failure rate, under single use environment, is calculated by using the following equation [9]:

$$\lambda_E = \sum_{j=1}^{n} q_j(\lambda_g \theta_f)_j \tag{3.38}$$

where

n = total number of different generic component/part classifications in the equipment
λ_E = total equipment failure rate, expressed in failures/10^6 h
θ_f = quality factor of generic component j
λ_g = generic failure rate of generic component j, expressed in failures/10^6 h
q_j = quantity of generic component j

Values for θ_f and λ_g are tabulated in Ref. [9].

Example 3.4

Failure rate of an elapsed time meter is expressed by [9]

$$\lambda_{EM} = \lambda_b f_e f_t \quad \text{failures}/10^6\,\text{h} \tag{3.39}$$

where λ_{EM} is the elapsed time meter failure rate and f_t is the temperature stress factor.

The following data are available for the meter [9]:

- $\lambda_b = 20$ failures/10^6 h
- $f_e = 2$ (ground, fixed)
- $f_t = 0.6$

Calculate the elapsed time meter failure rate.
Substituting the given data into Equation (3.39), we get

$$\lambda_{EM} = (20)(2)(0.6)$$
$$= 24 \text{ failures} / 10^6 \text{ h}$$

Thus, the elapsed time meter failure rate is 24 failures/10^6 h.

3.7 Failure Data Collection, Sources, and Failure Rates for Selected Items

Failure data provide invaluable information to professionals such as reliability engineers, design engineers, and system engineers concerning product performance in the field. In fact, it may be added that failure data are the backbone of reliability analysis because times to failure distribution of a product/part can only be established from its past failure history.

Over the years, many different ways and means have been used to collect failure data. Nonetheless, in the equipment/product life cycle, there are basically eight identifiable sources for collecting failure-related data as shown in Figure 3.2 [10].

Some of the specific/principal sources/documents for obtaining failure-related data are as follows:

- Government industry data exchange program [11].
- American Institute for Research (AIR) Data Store [12].
- Nuclear plant reliability data system [13].
- MIL-HDBK-217, Reliability prediction of electronic equipment [9].
- IEEE-STD-500-1977, IEEE Nuclear reliability data [14].
- SYREL: Reliability data bank [15].
- European Space Agency (ESA) Electronic Components Data Bank [16].
- Data on general equipment [17].
- NPRD-3 (Report), Nonelectronic parts reliability data [18].
- Book: Mechanical reliability: theory, models, and applications (it contains some data and lists over 55 failure data sources) [19].

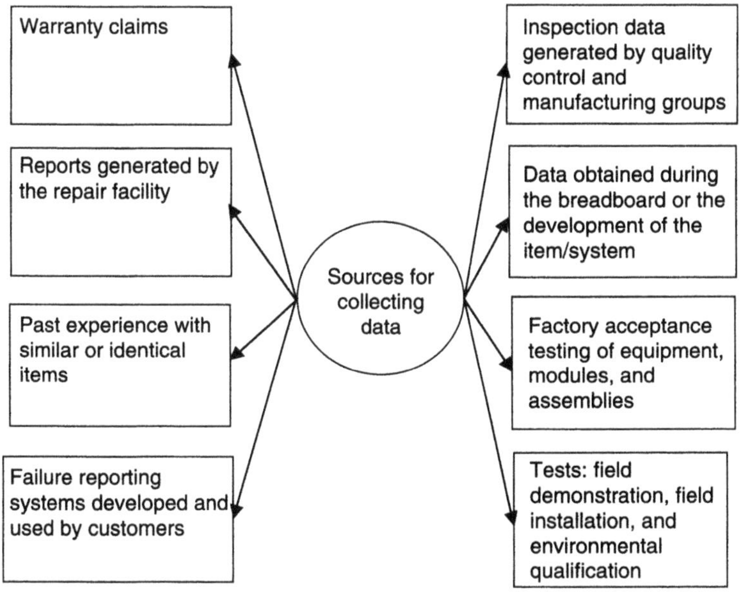

FIGURE 3.2
Identifiable sources for collecting failure-related data during the product life cycle.

- Book: Human reliability: with human factors (it contains some data and lists over 24 human reliability data sources) [20].
- Book: Robot reliability and safety (it contains some data and lists over 55 failure-related data sources) [21].

3.7.1 Failure Rates for Selected Items

In the published literature a large amount of failure rate data are available. Table 3.1 presents failure rates for selective items taken from the published data sources [4, 9, 18–19, 22].

3.8 Problems

1. Discuss the need for reliability.
2. Draw the bathtub hazard rate curve and discuss its three important regions.
3. Obtain a hazard rate expression for Weibull distribution by using its probability density function.
4. Obtain a reliability expression for Weibull distribution by using its hazard rate function.

TABLE 3.1

Failure Rates for Selected Items

No.	Item	Failure rate/10^6 h
1	Neon lamp	0.20
2	Lamp, incandescent	1
3	Single fiber optic connectors	0.10
4	Washer	0.5
5	Knob	0.09
6	Motor, general	10.00
7	Seal (O-ring)	0.2
8	Seal (sliding)	3.0
9	Hair spring	1
10	Nut or bolt	0.02
11	Commercial grade hybrid relay (use environment: ground, benign)	0.0551
12	Crimp connection (use environment: ground, benign)	0.00026
13	Vibrator (MIL-V-95): 400-cycle	40
14	Fiber optic cables (single fiber types only)	0.1 (per fiber Km)
15	Relief valve	0.5–10

5. Prove that reliability, $R(t)$ is given by

$$R(t) = \exp\left[-\int_0^t \lambda(t)dt\right] \tag{3.40}$$

where t is time and $\lambda(t)$ is the hazard or time-dependent failure rate.
6. Compare hazard rates of exponential and Rayleigh distributions.
7. Write down three different formulas for obtaining mean time to failure.
8. What are the important sources for collecting failure-related data during the product life cycle.
9. Prove that mean time to failure is given by

$$\text{MTTF} = \lim_{s\to 0} R(s) \tag{3.41}$$

where

MTTF = mean time to failure
s = Laplace transform variable
$R(s)$ = Laplace transform of the reliability function, $R(t)$

10. Define gamma function.
11. Obtain a mean time to failure expression using Equation (3.29) and Equation (3.34).

References

1. Elsayed, E.A., *Reliability Engineering*, Addison-Wesley, Reading, MA, 1996.
2. Wilkie, T., Soviet engineers admit failings in reactor design, *New Scientist*, August 28, 1986, pp. 14–15.
3. *Report of the Presidential Commission on the Space Shuttle Challenger Accident*, Vol. 1, Washington, D.C., June 6, 1986.
4. Dhillon, B.S., *Design Reliability: Fundamentals and Applications*, Boca Raton, FL, 1999.
5. Kapur, K.C., Reliability and maintainability, in: *Handbook of Industrial Engineering*, edited by G. Salvendy, John Wiley & Sons, New York, 1982, pp. 8.5.1–8.5.34.
6. Shooman, M.L., *Probabilistic Reliability: An Engineering Approach*, McGraw-Hill, New York, 1968.
7. Dhillon, B.S., A hazard rate model, *IEEE Transactions on Reliability*, Vol. 29, 1979, pp. 150–151.
8. Spiegel, M.R., *Theory and Problems of Laplace Transforms*, McGraw-Hill, New York, 1965.
9. MIL-HDBK-217, *Reliability Prediction of Electronic Equipment*, Department of Defense, Washington, D.C.
10. Hahn, R.F., Data collection techniques, *Proc. Annu. Reliability and Maintainability Symp.* 1972, pp. 38–43.
11. Government Industrial Data Exchange Program (GIDEP), GIDEP Operations Center, Naval Weapons Station, Department of the Navy, Corona, CA.
12. Munger, S.J., Smith, R.W., and Payne, D., *An Index of Electronic Equipment Operability: Data Store*, Report. AIR-C43-1/62RP(1), American Institute for Research, Pittsburgh, 1962.
13. *Reporting Procedures Manual for the Nuclear Plant Reliability Data System*, Southwest Research Institute, San Antonio, TX, December 1980.
14. IEEE-STD-500-1977, *IEEE Nuclear Reliability Data Manual*, John Wiley & Sons, New York, 1977.
15. *Systems Reliability Service, Safety and Reliability Directorate*, UKAEA, Warrington, UK.
16. European Space Agency (ESA), Electronic Components Data Bank, Space Documentation Service, European Space Agency, via Galileo Galilei, Frascati, Italy.
17. Data on General Equipment, Military Electronics Laboratory, FACK, Stockholm, Sweden.
18. NPRD-3 (Report), Non-electronic Parts Reliability Data, Reliability Analysis Center, Rome Air Development Center, Griffiss Air Force Base, Rome, New York.
19. Dhillon, B.S., *Mechanical Reliability: Theory, Models, and Applications*, American Institute of Aeronautics and Astronautics, Washington, D.C., 1988.
20. Dhillon, B.S., *Human Reliability: With Human Factors*, Pergamon Press, New York, 1986.
21. Dhillon, B.S., *Robot Reliability and Safety*, Springer-Verlag, New York, 1991.
22. Green, A.E. and Bourne, A.J., *Reliability Technology*, John Wiley & Sons, New York, 1972.

4

Static Reliability Evaluation Models

4.1 Introduction

Many times reliability of engineering systems is evaluated without considering the time factor. More specifically, the reliability of system components is assumed constant. This type of evaluation or analysis is usually a form of preliminary or elementary analysis.

Nonetheless, in this type of reliability evaluation a system is represented by a block diagram. In turn, the block diagram is composed of many sub blocks representing subsystems/units/parts. The probability of failure or success of each of these subsystems, units, or parts is estimated to calculate the probability of failure of success of the overall system. In this case, the components' probability of success or failure does not vary with time (i.e., it remains constant).

Generally, in this type of reliability evaluation or analysis, it is assumed that components/units/subsystems fail independently. This chapter presents reliability evaluation of series, parallel, *k*-out-of-*m*, parallel-series, series-parallel, and bridge networks with constant units/parts/components reliabilities or failure probabilities.

4.2 Series Network

This network is the simplest of all reliability networks and its block diagram is shown in Figure 4.1. Each block in the diagram represents a unit/component. More specifically, the Figure 4.1 diagram represents a system with *m* number of units acting in series. If any one of the units fails, the system fails. In other words, all units must operate normally for the system's success.

The reliability of Figure 4.1 series systems network is expressed by [1]

$$R_S = P(x_1 x_2 x_3 \ldots x_m) \tag{4.1}$$

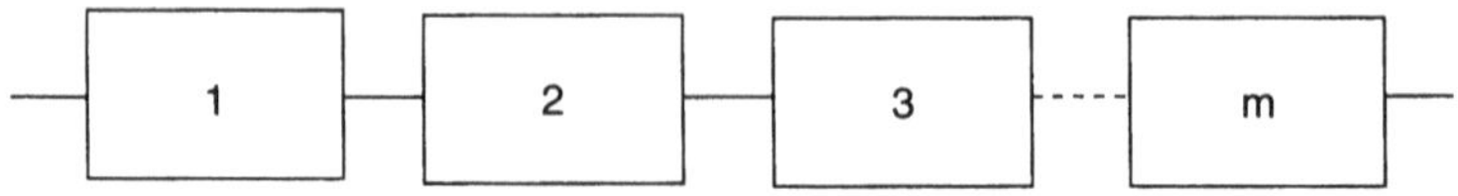

FIGURE 4.1
Series network.

where

R_S = series system reliability or probability of success
x_i = event denoting the success of unit i, for $i = 1, 2, 3, \ldots, m$
$P(x_1 x_2 x_3 \ldots x_m)$ = probability of occurrence of events $x_1, x_2, x_3, \ldots, x_m$

For independently failing units, Equation (4.1) becomes

$$R_S = P(x_1)P(x_2)P(x_3)\ldots P(x_m) \tag{4.2}$$

where $P(x_i)$ is the occurrence probability of event x_i, for $i = 1, 2, 3, \ldots, m$.

If we let $R_i = P(x_i)$ in Equation (4.2) it becomes

$$R_S = \prod_{i=1}^{m} R_i \tag{4.3}$$

where R_i is the unit i reliability, for $i = 1, 2, 3, \ldots, m$.

For $R_i > 0.95$ in Equation (4.3), system reliability R_S can be approximated by using equation

$$R_S = 1 - \sum_{i=1}^{m} (1 - R_i) \tag{4.4}$$

For identical units (i.e., $R_i = R$) Equation (4.4) becomes

$$R_S = 1 - m(1 - R) \tag{4.5}$$

where R is the unit reliability.

Example 4.1

Assume that an automobile has four independent and identical tires. The tire reliability is 0.97. If any one of the tires is punctured, the automobile cannot be driven. Calculate the automobile reliability with respect to tires by using Equation (4.3) and Equation (4.5). Comment on the end result.

By substituting the given data into Equation (4.3), we get

$$\begin{aligned} R_S &= (0.97)(0.97)(0.97)(0.97) \\ &= 0.8853 \end{aligned}$$

Similarly, using the given data values in Equation (4.5) yields

$$R_S = 1 - 4(1 - 0.97)$$
$$= 0.88$$

Both the above reliability results are very close. More specifically, the system reliability value obtained through using Equation (4.5) is lower than when the exact equation (i.e., Equation (4.3)) was used.

4.3 Parallel Network

This is a widely used network and it represents a system with m units operating simultaneously. At least one unit must operate normally for the system success. The m-unit parallel system block diagram is shown in Figure 4.2.

Each block in the diagram denotes a unit. The failure probability of the Figure 4.2 parallel system/network is given by [1, 3].

$$F_p = P(\bar{x}_1 \bar{x}_2 \bar{x}_3 \ldots \bar{x}_m) \tag{4.6}$$

where

F_p = failure probability of the parallel system
$\bar{x}_i$ = event denoting the failure of unit i; for $i = 1, 2, 3, \ldots, m$
$P(\bar{x}_1 \bar{x}_2 \bar{x}_3 \ldots \bar{x}_m)$ = probability of occurrence of events $\bar{x}_1, \bar{x}_2, \bar{x}_3, \ldots, \bar{x}_m$

For independently failing units, Equation (4.6) becomes

$$F_p = P(\bar{x}_1)P(\bar{x}_2)P(\bar{x}_3)\ldots P(\bar{x}_m) \tag{4.7}$$

where $P(\bar{x}_i)$ is the probability of occurrence of failure event $\bar{x}_i$, for $i = 1, 2, 3, \ldots, m$.

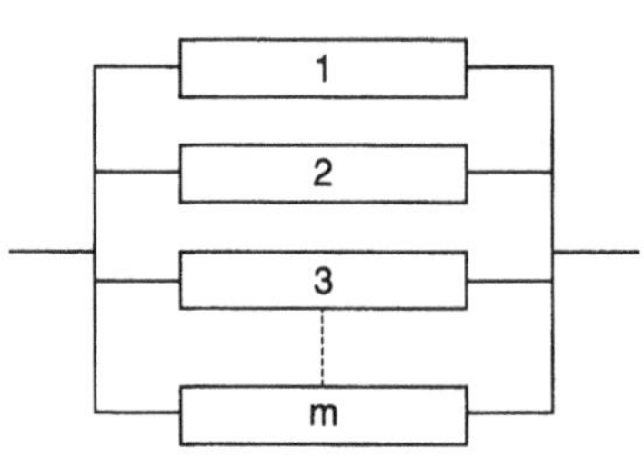

FIGURE 4.2
Parallel network.

If we let $F_i = P(x_i)$ in Equation (4.7) it yields

$$F_p = \prod_{i=1}^{m} F_i \tag{4.8}$$

where F_i is the failure probability of unit i for $i = 1, 2, 3, \ldots, m$.

Subtracting Equation (4.8) from unity yields the following expression for parallel network reliability:

$$\begin{aligned} R_p &= 1 - F_p \\ &= 1 - \prod_{i=1}^{m} F_i \end{aligned} \tag{4.9}$$

where R_p is the parallel system reliability.

For identical units, Equation (4.9) becomes

$$R_p = 1 - F^m \tag{4.10}$$

where F is the unit failure probability.

Since $R + F = 1$, Equation (4.10) is rewritten to the following form:

$$R_p = 1 - (1 - R)^m \tag{4.11}$$

where R is the unit reliability.

The plots of Equation (4.11) are shown in Figure 4.3. The figure clearly demonstrates that as the unit reliability and the number of redundant units increase, the parallel system reliability increases accordingly.

Example 4.2

A computer has two independent and identical Central Processing Units (CPUs) operating simultaneously. At least one CPU must operate normally for the computer to function successfully. If the CPU reliability is 0.96, calculate the computer reliability with respect to CPUs.

By substituting the specified data values into Equation (4.11), we get

$$\begin{aligned} R_p &= 1 - (1 - 0.96)^2 \\ &= 0.9984 \end{aligned}$$

Thus, the computer reliability with respect to CPUs is 0.9984.

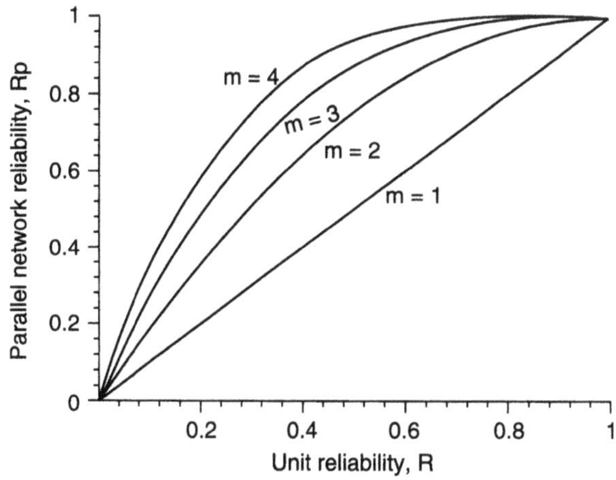

FIGURE 4.3
Parallel system reliability plots.

4.4 *k*-Out-of-*m* Network

This is another form of redundancy and is frequently used in computers. In this case, at least k units out of a total of m units must operate normally for the system success. Just like in the case of the parallel network, all the units in this network are active or operating simultaneously.

For independent and identical units, using the Binomial distribution, we get the following expression for the k-out-of-m unit system reliability [2–3]:

$$R_{k/m} = \sum_{i=k}^{m} \binom{m}{i} R^i (1-R)^{m-i} \tag{4.12}$$

where

$\binom{m}{i} = \frac{mi}{(m-i)!\,i!}$

$R_{k/m}$ = k-out-of-m network/system reliability

R = unit reliability

For $k = m$, Equation (4.12) is the same as the one for the series network and for $k = 1$ Equation (4.12) becomes same for identical units parallel network. In other words, series and parallel networks are the special cases of this network.

The plots of Equation (4.12) for $m = 1$, (i.e., single unit system); $k = 3$, $m = 3$ (i.e., three unit series system); $k = 1$, $m = 3$ (i.e., three unit parallel system); and $k = 2$, $m = 3$ (i.e., 2-out-of-3 units system) are shown in Figure 4.4.

These plots demonstrate that the highest system reliability is achieved when only one unit out of three units is needed for the system success (i.e., three

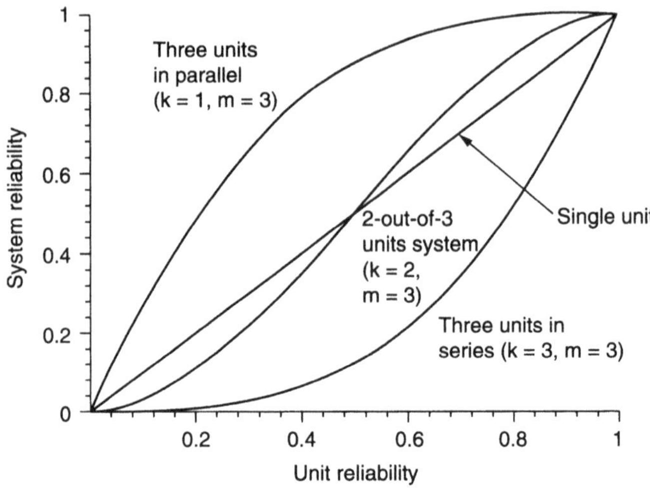

FIGURE 4.4
k-out-of-*m* units system reliability plots.

unit parallel system). Reliability expressions for a *k*-out-of-*m* units network with nonidentical units may be obtained by using the Binomial formula.

Example 4.3

Assume that an aircraft has three identical and independent engines. At least two engines must operate normally for aircraft to fly successfully. The engine reliability is 0.97. Calculate the reliability of the aircraft with respect to engines. By substituting the specified data into Equation (4.12), we get

$$R_{2/3} = \sum_{i=2}^{3} \binom{3}{i} R^i (1-R)^{3-i}$$

$$= 3R^2 - 2R^3$$

$$= 3(0.97)^2 - 2(0.97)^3$$

$$= 0.9974$$

Thus, the reliability of the aircraft with respect to engines is 0.9974.

4.5 Series-Parallel Network

This network represents a system having *m* number of subsystems in series. In turn, each subsystem contains *k* number of active (i.e., operating) units in parallel. If any one of the subsystems fails, the system fails. The network/system

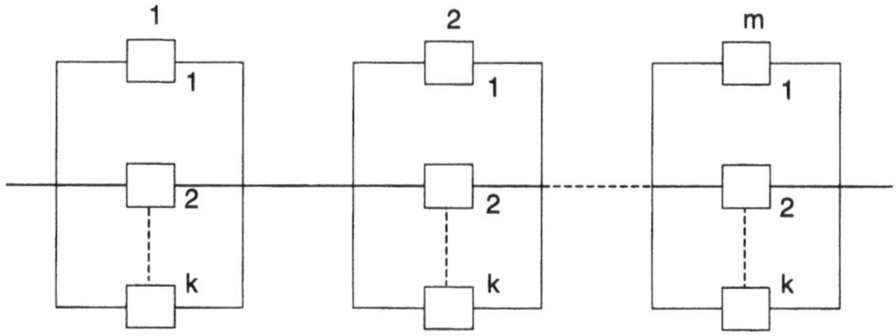

FIGURE 4.5
Series-parallel system/network.

block diagram is shown in Figure 4.5. Each block in the diagram represents a unit.

For independent units, using Equation (4.9) we write the following equation for the ith Figure 4.5 subsystem's reliability [3, 4]

$$R_{pi} = 1 - \prod_{j=1}^{k} F_{ij} \tag{4.13}$$

where R_{pi} is the reliability of the parallel subsystem i and F_{ij} is the ith subsystem's jth unit's failure probability.

Substituting Equation (4.13) into Equation (4.3) yields the following expression for series-parallel network/system reliability:

$$R_{sp} = \prod_{i=1}^{m} \left(1 - \prod_{j=1}^{k} F_{ij} \right) \tag{4.14}$$

where R_{sp} is the series-parallel network/system reliability.

For identical units, Equation (4.14) becomes

$$R_{sp} = (1 - F^k)^m \tag{4.15}$$

where F is the unit failure probability. Since $R + F = 1$, Equation (4.15) is rewritten to the following form:

$$R_{sp} = [1 - (1 - R)^k]^m \tag{4.16}$$

where R is the unit reliability. For $R = 0.8$, the plots of Equation (4.16) are shown in Figure 4.6. These plots indicate that as the number of subsystems m increases, the system reliability decreases, accordingly. On the other hand, as the number of units k increases, the system reliability also increases.

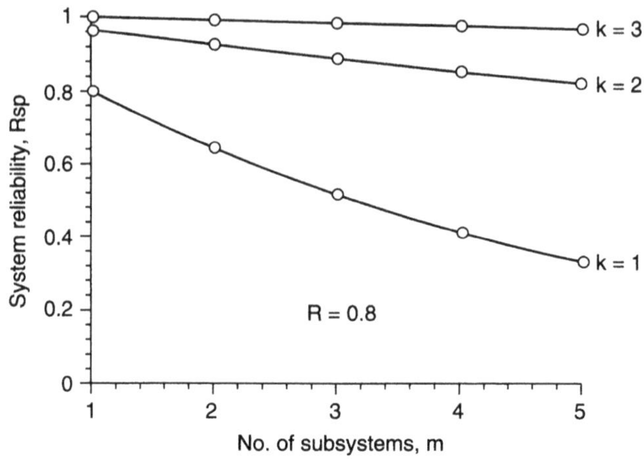

FIGURE 4.6
Series-parallel system reliability plots.

Example 4.4

Assume that a system has four active, independent, and identical units forming a series-parallel configuration (i.e., $k=2$, $m=2$). Each unit's reliability is 0.94. Calculate the system reliability.

By substituting the given data values into Equation (4.16) yields

$$R_{sp} = [1-(1-0.94)^2]^2$$
$$= 0.9928$$

Thus, the system reliability is 0.9928.

4.6 Parallel-Series Network

This network represents a system having m number of subsystems in parallel. In turn, each subsystem contains k number of active (i.e., operating) units in series. At least one subsystem must function normally for the system success. The network/system block diagram is shown in Figure 4.7. Each block in the diagram denotes a unit.

For independent and identical units, using Equation (4.3), we get the following equation for the ith Figure 4.7 subsystem's reliability:

$$R_{si} = \prod_{j=1}^{k} R_{ij} \tag{4.17}$$

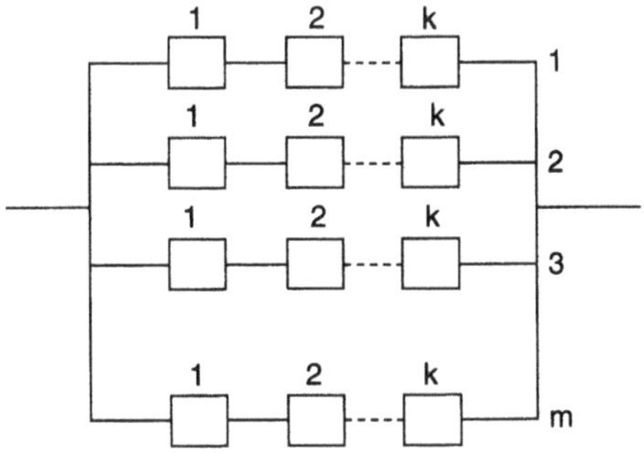

FIGURE 4.7
Parallel-series network system.

where R_{si} is the reliability of the series subsystem i and R_{ij} is the ith subsystem's jth unit's reliability. By subtracting Equation (4.17) from unity, we get

$$F_{si} = 1 - R_{si} = 1 - \prod_{j=1}^{k} R_{ij} \tag{4.18}$$

where F_{si} is the failure probability of the series subsystem i.

Using Equation (4.18) in Equation (4.9) yields

$$R_{ps} = 1 - \prod_{i=1}^{m} \left(1 - \prod_{j=1}^{k} R_{ij} \right) \tag{4.19}$$

where R_{ps} is the parallel-series network/system reliability. For identical units Equation (4.19) simplifies to

$$R_{ps} = 1 - (1 - R^k)^m \tag{4.20}$$

where R is the unit reliability.

For $R = 0.8$, the plots of Equation (4.20) are shown in Figure 4.8. The plots show that as the number of units k increases, the system/network reliability decreases accordingly. On the other hand, as the number of subsystems m increases, the system reliability also increases.

Example 4.5

A system is composed of four active, independent, and identical units forming a parallel-series configuration (i.e., $k = m = 2$). Calculate the system reliability, if each unit's reliability is 0.94.

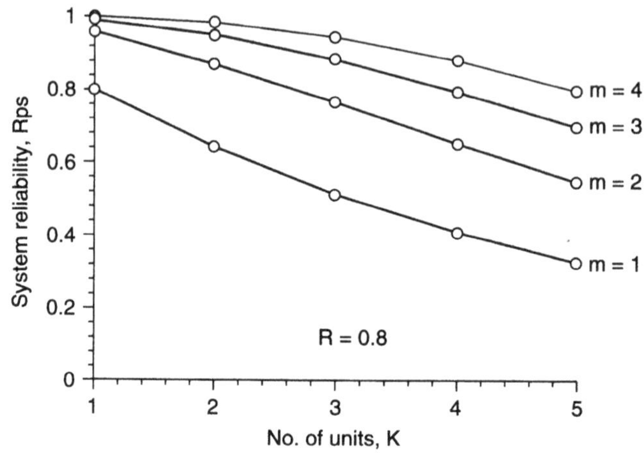

FIGURE 4.8
Parallel-series system reliability plots.

By substituting the given data into Equation (4.20), we get

$$R_{ps} = 1 - [1 - (0.94)^2]^2$$
$$= 0.9865$$

Thus, parallel-series system reliability is 0.9865.

4.7 Bridge Network

This is another configuration that time to time occurs in engineering systems. The configuration's block diagram is shown in Figure 4.9. Each block in the diagram denotes a unit.

For independent units, the bridge network shown in Figure 4.9 reliability is expressed by [5]

$$\begin{aligned} R_b = {} & 2R_1R_2R_3R_4R_5 + R_2R_3R_4 + R_1R_3R_5 + R_1R_4 \\ & + R_2R_5 - R_2R_3R_4R_5 - R_1R_2R_3R_4 - R_5R_1R_2R_3 \\ & - R_1R_3R_4R_5 - R_1R_2R_4R_5 \end{aligned} \tag{4.21}$$

where R_b is the bridge network reliability and R_i is the unit i reliability, for $i = 1, 2, 3, 4$, and 5.

For identical units Equation (4.21) becomes

$$R_b = 2R^5 - 5R^4 + 2R^3 + 2R^2 \tag{4.22}$$

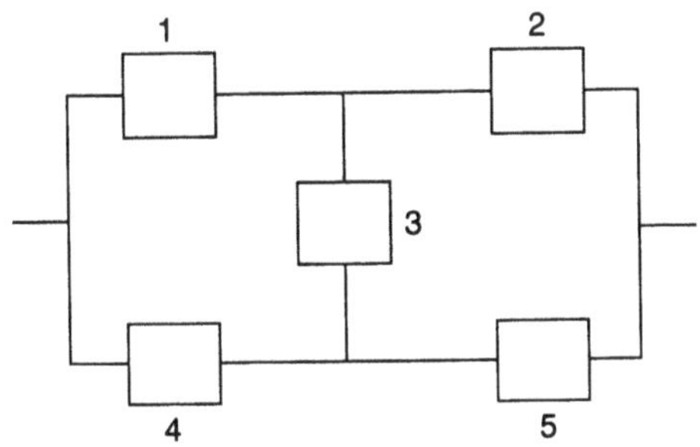

FIGURE 4.9
Bridge network

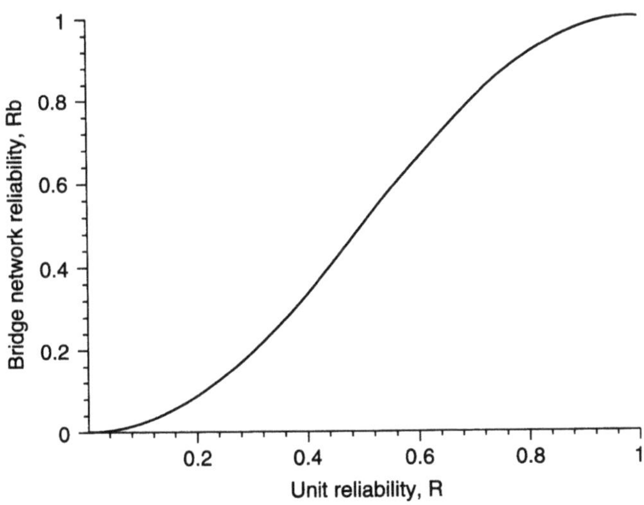

FIGURE 4.10
Bridge network reliability plot.

where R is the unit reliability. The plot of Equation (4.22) is shown in Figure 4.10.

Example 4.6

A communication subsystem is composed of five independent and identical units forming a bridge network. Calculate the network reliability, if each unit's reliability is 0.98.

By substituting the specified data into Equation (4.22) yields

$$R_b = 2(0.98)^5 - 5(0.98)^4 + 2(0.98)^3 + 2(0.98)^2$$
$$= 0.9992$$

Thus, the bridge network's reliability is 0.9992.

4.8 Problems

1. Compare series and parallel networks.
2. What are the special case networks of the k-out-of-m units systems?
3. Compare series-parallel and parallel-series networks.
4. Prove that the reliability of a parallel network is given by

$$R_p = 1 - \prod_{i=1}^{m} (1 - R_i) \tag{4.23}$$

where

R_p = parallel network reliability
m = total number of units in parallel
R_i = reliability of unit i, for $i = 1, 2, 3, \ldots, m$

5. Prove that the reliability of a parallel-series network is expressed by

$$R_{sp} = [1 - (1 - R)^k]^m \tag{4.24}$$

where

k = number units in parallel
m = number of subsystems
R = unit reliability

6. Compare a five-unit bridge network with a five-unit parallel network.
7. A system has three independent, identical, and active units. At least two units must operate normally for the system success. The reliability of each unit is 0.91. Calculate the system reliability.
8. An aircraft has four active, independent, and identical engines. At 35000 ft above ground at least one engine must operate normally for the aircraft to fly successfully. Calculate the reliability of the aircraft flying at 35000 ft, if the engine probability of failure is 0.05.
9. Prove that the reliability of a series system is given by

$$R_s = 1 - \sum_{i=1}^{m} (1 - R_i) \tag{4.25}$$

where

R_s = series system reliability
m = total number of units in series
R_i = reliability of unit i, for $i = 1, 2, 3, \ldots, m$

10. Assume that an automobile has four independent and identical tires. The tire reliability is 0.93. If any one of the tires is punctured, the automobile cannot be driven. Calculate the automobile reliability with respect to tires.

References

1. Shooman, M.L., *Probabilistic Reliability: An Engineering Approach,* McGraw-Hill, New York, 1968.
2. AMCP 702-3, *Quality Assurance: Reliability Handbook,* Headquarters, U.S. Army Material Command, Washington, D.C., 1968.
3. Dhillon, B.S., *Systems Reliability, Maintainability, and Management,* Petrocelli Books, New York, 1983.
4. Von Alven, W.H., Ed., *Reliability Engineering,* Prentice Hall, Englewood Cliffs, NJ, 1964.
5. Lipp, J.P., Topology of switching elements versus reliability, *Trans. IRE Reliability and Quality Control,* Vol. 7, 1957, pp. 21–34.

5

Dynamic Reliability Evaluation Models

5.1 Introduction

Usually, in the real life environment failure data-related information on engineering items is collected and the values of parameters such as mean time to failure and failure rates are computed. Under such scenario, the reliability of items varies with time. More specifically, the item's reliability is nonconstant.

The time to failure of items can follow various distinct patterns. Over the years, many different statistical distributions have been used to represent these patterns. These distributions include exponential, Weibull, normal, gamma, and Rayleigh [1–4]. Past experiences indicate that generally the time to failure of electronic parts follows the exponential distribution and the mechanical parts the Weibull. Nonetheless, in the real life reliability evaluation of engineering systems/parts, the exponential distribution is most widely used. Probably the most important factor for its wide spread application is its simplicity to use.

This chapter is concerned with reliability evaluation of various standard reliability configuration under dynamic conditions (i.e., when the item failure probability varies with time).

5.2 Series Network

For time t dependent unit reliability, the series network/system reliability from Equation (4.3) is

$$R_s(t) = \prod_{i=1}^{m} R_i(t) \tag{5.1}$$

where

$R_s(t)$ = series network/system reliability at time t

m = number of units in series

$R_i(t)$ = unit i reliability at time t, for $i = 1, 2, 3, \ldots, m$

For exponentially distributed times to failure (i.e., constant failure rate) of unit i, the unit reliability is expressed by [5].

$$R_i(t) = e^{-\lambda_i t} \tag{5.2}$$

where $R_i(t)$ is the ith unit reliability at time t and λ_i is the constant failure rate of unit i.

By substituting Equation (5.2) into Equation (5.1), we get

$$R_s(t) = e^{-\sum_{i=1}^{m} \lambda_i t} \tag{5.3}$$

For $\lambda_i t \ll 1$, Equation (5.3) becomes

$$R_s(t) \approx 1 - \sum_{i=1}^{m} \lambda_i t \tag{5.4}$$

Past experiences indicate that Equation (5.4) yields good approximation when $\lambda_i t \le 0.05$.

By integrating Equation (5.3) over the time interval $[0, \infty]$, we get the following equation for the series system mean time to failure [5].

$$\begin{aligned} \text{MTTF}_s &= \int_0^{\infty} e^{-\sum_{i=1}^{m} \lambda_i t} dt \\ &= \frac{1}{\sum_{i=1}^{m} \lambda_i} \end{aligned} \tag{5.5}$$

where MTTF_s is the series system mean time to failure.

The series system hazard rate $\lambda_s(t)$ is given by [2]

$$\begin{aligned} \lambda_s(t) &= \frac{f_s(t)}{R_s(t)} = -\frac{1}{R_s(t)} \frac{dR_s(t)}{dt} \\ &= -\frac{1}{e^{-\sum_{i=1}^{m} \lambda_i t}} \left(-\sum_{i=1}^{m} \lambda_i \right) e^{-\sum_{i=1}^{m} \lambda_i t} \\ &= \sum_{i=1}^{m} \lambda_i \end{aligned} \tag{5.6}$$

where $f_s(t)$ is the failure or probability density function of the series system.

Equation (5.6) demonstrates that when one adds failure rates of items, he/she automatically assumes that those items are connected in series.

Example 5.1

Assume that in Example 5.1 the tire failure rate λ_t is given (i.e., $\lambda_t = 0.0004$ failures/hour) instead of reliability. Calculate the automobile failure rate and mean time to failure with respect to tires.

For identical tires by substituting the given data value into Equation (5.6) we get

$$\lambda_s = 4(0.0004)$$
$$= 0.0016 \text{ failures/h}$$

Using the above result in Equation (5.5) yields

$$MTTF_s = 625 \text{ h}$$

Thus, the automobile failure rate and mean time to failure with respect to tires are 0.0016 failures/h and 625 h, respectively.

5.2.1 Series System Reliability and Mean Time to Failure with Weibull Distributed Units' Times to Failure

When an *i*th unit's times to failure are Weibull distributed, its reliability is expressed by [2]

$$R_i(t) = e^{-(\theta_i t^{K+1})/(K+1)} \tag{5.7}$$

where

$R_i(t)$ = *i*th unit's reliability at time t

θ_i = *i*th unit's scale parameter

K = *i*th unit's shape parameter

By substituting Equation (5.7) into Equation (5.1) yields the following expression for the series system reliability:

$$R_s(t) = \prod_{i=1}^{m} e^{-(\theta_i t^{K+1})/(K+1)} \tag{5.8}$$

By integrating Equation (5.8) over the time interval [0, ∞] we get the following equation for the series system mean time to failure [2]:

$$\text{MTTF}_s = \Gamma\left(\frac{1}{K+1}\right)\left[(K+1)\left\{\frac{\sum_{i=1}^{m}\theta_i}{K+1}\right\}^{1/K+1}\right]^{-1} \tag{5.9}$$

where $\Gamma\left(\frac{1}{K+1}\right)$ is the gamma function and MTTF_s is the series system mean time to failure with Weibull distributed units' times to failure.

5.2.2 Series System Reliability with Nonconstant Failure Rates of Units

In this case, we assume that hazard or time dependent failure rate of the *i*th series unit is defined by [2, 6]

$$\lambda_i\,(t) = \lambda_i + \alpha_i\, t^{\theta} \tag{5.10}$$

where t is time and λ_i, θ, and α_i are the constants or parameters.

A general expression for obtaining an *i*th unit or item reliability is as follows [5]:

$$R_i(t) = e^{-\int_0^t \lambda_i(t)dt} \tag{5.11}$$

where $R_i\,(t)$ is the *i*th unit reliability at time t and $\lambda_i\,(t)$ is the *i*th unit hazard or time dependent failure rate.

Using Equation (5.10) in Equation (5.11) yields

$$R_i(t) = e^{-[\lambda_i t + \{\alpha_i t^{\theta+1}\}/(\theta+1)]} \tag{5.12}$$

By substituting Equation (5.12) into Equation (5.1) we get

$$R_s(t) = \exp\left[-\left\{\sum_{i=1}^{m}\lambda_i t + \left(t^{\theta+1}\sum_{i=1}^{m}\alpha_i\right)\Big/(\theta+1)\right\}\right] \tag{5.13}$$

Now let

$$\lambda_{tx} = \sum_{i=1}^{m}\lambda_i \tag{5.14}$$

$$T = t\,\lambda_{tx} \tag{5.15}$$

$$\alpha_{tx} = \sum_{i=1}^{m}\alpha_i \tag{5.16}$$

Thus, Equation (5.13) becomes

$$R_s(t) = \exp\left[-\left\{T + \frac{\alpha_{tx} T^{(\theta+1)}}{\lambda_{tx}(\theta+1)\lambda_{tx}^{\theta}}\right\}\right] \tag{5.17}$$

For a very large value of m, we assume

$$\lambda_{tx} \to \infty,$$
$$\{\alpha_{tx}/(\theta+1)\lambda_{tx}\} \text{ is bounded.}$$

Thus, for a very large value of m, Equation (5.17) yields the following expression for the series system reliability:

$$\begin{aligned} R_s &= e^{-T} \\ &= e^{-(\lambda_{tx})t} \\ &= e^{-\left(\sum_{i=1}^{m} \lambda_i\right)t} \end{aligned} \tag{5.18}$$

Equation (5.18) demonstrates that for a large number of units or components with non-constant failure rates, the series system reliability is the same as for the series system with constant unit or component failure rates. More specifically, Equation (5.3) and Equation (5.18) are identical.

5.3 Parallel Network

For time t dependent unit reliability, the parallel network/system reliability from Equation (4.9) is

$$R_p(t) = 1 - \prod_{i=1}^{m} F_i(t) \tag{5.19}$$

where

m = number of units in parallel

$R_p(t)$ = parallel network/system reliability at t

$F_i(t) = 1 - R_i(t)$ = time-dependent failure probability of unit i, for $i = 1, 2, 3, \ldots, m$

$R_i(t)$ = time-dependent reliability of unit i, for $i = 1, 2, 3, \ldots, m$

Since $F_i(t) = 1 - R_i(t)$, Equation (5.19) becomes

$$R_p(t) = 1 - \prod_{i=1}^{m} (1 - R_i(t)) \tag{5.20}$$

For constant failure rate or exponentially distributed times to failure of unit i, by substituting Equation (5.2) into Equation (5.20) we get

$$R_p(t) = 1 - \prod_{i=1}^{m} (1 - e^{-\lambda_i t}) \tag{5.21}$$

For identical units, Equation (5.21) becomes

$$R_p(t) = 1 - (1 - e^{-\lambda t})^m \tag{5.22}$$

where λ is the unit constant rate. For $\lambda_i\, t \leq 0.05$, Equation (5.21) yields

$$R_p(t) \approx 1 - \prod_{i=1}^{m} \lambda_i t^m \tag{5.23}$$

For identical units, Equation (5.23) simplifies to

$$R_p\ (t) \approx 1 - (\lambda t)^m \tag{5.24}$$

For the specified values of m, the plots of Equation (5.22) are shown in Figure 5.1.

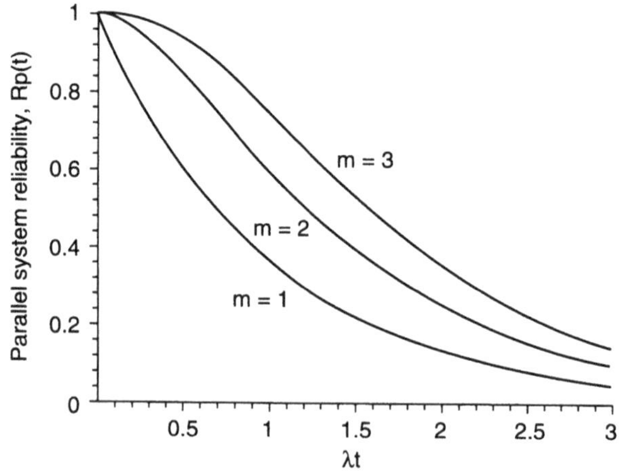

FIGURE 5.1
Parallel system time-dependent reliability plots.

By integrating Equation (5.21) over the time interval [*0*, ∞], we get the following expression for the parallel system mean time to failure [2]:

$$\text{MTTF}_p = \int_0^\infty \left[1 - \prod_{i=1}^{m}(1-e^{-\lambda_i t})\right]dt$$
$$= \left(\frac{1}{\lambda_1}+\frac{1}{\lambda_2}+\cdots+\frac{1}{\lambda_m}\right) - \left(\frac{1}{\lambda_1+\lambda_2}+\frac{1}{\lambda_2+\lambda_3}+\cdots+\frac{1}{\lambda_i+\lambda_j}\right)$$
$$+\left(\frac{1}{\lambda_1+\lambda_2+\lambda_3}+\frac{1}{\lambda_1+\lambda_2+\lambda_4}+\cdots+\frac{1}{\lambda_i+\lambda_j+\lambda_k}\right) - \cdots + \{(-1)^{m+1}\}\Big/\sum_{i=1}^{m}\lambda_i \tag{5.25}$$

where MTTF_p is the parallel network mean time to failure. For identical units, Equation (5.25) simplifies to

$$\text{MTTF}_p = \frac{1}{\lambda}\sum_{i=1}^{m}\frac{1}{i} \tag{5.26}$$

Example 5.2

A computer has two independent and identical central processing units operating in parallel. The constant failure rate of the central processing unit is 0.0002 failures/h. Calculate the computer mean time to failure with respect to central processing units.

By inserting the given data into Equation (5.26), we get

$$\text{MTTF}_p = \frac{1}{0.0002}\left(1+\frac{1}{2}\right)$$
$$= 7500\,\text{h}$$

Thus, the computer mean time to failure with respect to central processing units is 7500 h.

5.4 *k*-Out-of-*m* Network

For time *t* dependent unit reliability, the *k*-out-of-*m* network/system reliability from Equation (4.12) is

$$R_{K/m}(t) = \sum_{i=K}^{m}\binom{m}{i}[R(t)]^i[1-R(t)]^{m-i} \tag{5.27}$$

where

m = number of identical units
$R_{K/m}(t)$ = k-out-of-m unit network/system reliability at time t
$R(t)$ = unit reliability at time t

$$\binom{m}{i} = \frac{m!}{(m-i)!i!}$$

For constant failure rate or exponentially distributed times to failure of a unit, by inserting Equation (5.2) into Equation (5.27) yields

$$R_{K/m}(t) = \sum_{i=K}^{m} \binom{m}{i} e^{-i\lambda t} [1 - e^{-\lambda t}]^{m-i} \tag{5.28}$$

where λ is the unit failure rate.

For some given values of K/m, approximate expressions of Equation (5.28) are presented in Table 5.1 [7].

The plots of Equation (5.28) for $K = 2$, $m = 3$ (i.e., 2-out-of-3 units system); $m = 1$ (i.e., single unit system); $K = 1$, $m = 3$ (i.e., three-units parallel system); and $K = 3$, $m = 3$ (i.e., three-units series system) are shown in Figure. 5.2.

TABLE 5.1

Approximate Expressions of Equation (5.28) for Some Specified Values of K/m

No.	Specified K/m Values	Approximate Expression for Calculating System Reliability
1	$K = 1$, $m = 2$ (i.e., at least 1 unit out-of-2 must work for success)	$1-(\lambda t)^2$
2	$K = 1$, $m = 3$ (i.e., at least 1 unit out-of-3 must work for success)	$1-(\lambda t)^3$
3	$K = 1$, $m = m$ (i.e., at least 1 unit out-of-m must work for success)	$1-(\lambda t)^m$
4	$K = 2$, $m = 3$ (i.e., at least 2 units out-of-3 must work for success)	$1 - 3(\lambda t)^2$
5	$K = 3$, $m = 4$ (i.e., at least 3 units out-of-4 must work for success)	$1 - 6(\lambda t)^2$
6	$K = m - 1$, $m = m$ (i.e., at least $(m - 1)$ units out-of-m must work for success)	$1-\binom{m}{2}(\lambda t)^2$
7	$K = 2$, $m = 4$ (i.e., at least 2 units out-of-4 must work for success)	$1 - 4(\lambda t)^3$
8	$K = 3$, $m = 5$ (i.e., at least 3 units out-of-5 must work for success)	$1 - 10(\lambda t)^3$
9	$K = m - 2$, $m = m$ (i.e., at least $(m - 2)$ units out-of-m must work for success)	$1-\binom{m}{3}(\lambda t)^3$

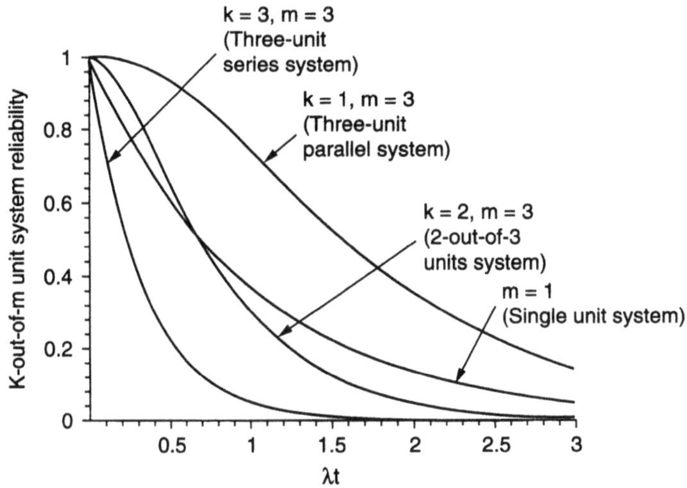

FIGURE. 5.2
k-out-of-*m* unit system time-dependent reliability plots.

By integrating Equation (5.28) over the time interval $[0, \infty]$, we get the following equation for *k*-out-of-*m* unit system mean time to failure:

$$\begin{aligned} \mathrm{MTTF}_{m/K} &= \int_0^\infty \left[\sum_{i=K}^{m} \binom{m}{i} e^{-i\lambda t} \{1 - e^{-\lambda t}\}^{m-i} \right] dt \\ &= \frac{1}{\lambda} \sum_{i=K}^{m} \frac{1}{i} \end{aligned} \tag{5.29}$$

where $\mathrm{MTTF}_{m/K}$ is the *k*-out-of-*m* unit system mean time to failure.

Example 5.3

An engineering system is composed of three active, independent, and identical units. At least two units must operate normally for the successful operation of the system. The unit constant failure rate is 0.0001 failures/h. Calculate the engineering system mean time to failure.

Using the given data in Equation (5.29) yields

$$\begin{aligned} \mathrm{MTTF}_{m/K} &= \frac{1}{(0.0001)} \left[\frac{1}{2} + \frac{1}{3} \right] \\ &= 8{,}333.3 \text{ h} \end{aligned}$$

Thus, the engineering system mean time to failure is 8333.3 h.

5.5 Series-Parallel Network

For time t dependent unit reliability, the series-parallel network/system reliability from Equation (4.14) is

$$R_{sp}(t) = \prod_{i=1}^{m}\left(1 - \prod_{j=1}^{K} F_{ij}(t)\right) \tag{5.30}$$

where

$R_{sp}(t)$ = series-parallel system reliability at time t
m = number of subsystems in series
K = number of active units in each parallel subsystem
$F_{ij}(t)$ = ith subsystem's jth unit's failure probability at time t

For constant failure rate or exponentially distributed time to failure of a unit ij, by subtracting Equation (5.2) from unity we get

$$\begin{aligned} F_{ij}(t) &= 1 - R_{ij}(t) \\ &= 1 - e^{-\lambda_{ij} t} \end{aligned} \tag{5.31}$$

where $F_i(t)$ is the ith/jth unit failure probability at time t and $R_{ij}(t)$ is the reliability of unit ij. Substituting Equation (5.31) into Equation (5.30) yields

$$R_{sp}(t) = \prod_{i=1}^{m}\left(1 - \prod_{j=1}^{K} (1 - e^{-\lambda_{ij} t})\right) \tag{5.32}$$

For identical units, Equation (5.32) becomes

$$R_{sp}(t) = [1 - (1 - e^{-\lambda t})^K]^m \tag{5.33}$$

where λ is the unit constant failure rate. By integrating Equation (5.33) over the time interval [0, ∞], we get the following equation for series-parallel network/system mean time to failure:

$$\begin{aligned} \text{MTTF}_{sp} &= \int_0^{\infty} [\{1 - (1 - e^{-\lambda t})^K\}^m]\, dt \\ &= \frac{1}{\lambda}\sum_{i=1}^{m}\left[(-1)^{i+1}\binom{m}{i}\sum_{j=1}^{iK}\frac{1}{j}\right] \end{aligned} \tag{5.34}$$

where MTTF_{sp} is the series-parallel network mean time to failure.

Example 5.4

A system contains four active, independent, and identical units forming a series-parallel configuration. The constant failure rate of each unit is 0.0004 failures/h. Calculate the system mean time to failure, if each parallel subsystem contains two units.

By substituting the specified given data values into Equation (5.34) yields

$$MTTF_{sp} = \frac{1}{(0.0004)} \sum_{i=1}^{2} \left[(-1)^{i+1} \binom{2}{i} \sum_{j=1}^{i2} \frac{1}{j} \right]$$

$$= 2291.7\text{h}$$

Thus, the system mean time to failure is 2291.7 h.

5.6 Parallel-Series Network

For time t dependent unit reliability, the parallel-series network/system reliability from Equation (4.19) is

$$R_{ps}(t) = 1 - \prod_{i=1}^{m} \left(1 - \prod_{j=1}^{K} R_{ij}(t) \right) \tag{5.35}$$

where

$R_{ps}(t)$ = parallel-series network/system reliability at time t
m = number of subsystems in parallel
K = number of active units in series in each subsystem
$R_{ij}(t)$ = ith subsystem's jth unit's reliability at time t

For constant failure rate or exponentially distributed times to failure of unit ij, by substituting Equation (5.2) into Equation (5.35), we get

$$R_{ps}(t) = 1 - \prod_{i=1}^{m} \left(1 - \prod_{j=1}^{K} e^{-\lambda_{ij} t} \right) \tag{5.36}$$

where λ_{ij} is the constant failure rate of unit ij.

For identical units, Equation (5.36) becomes

$$R_{ps}(t) = 1 - (1 - e^{-K\lambda t})^{m} \tag{5.37}$$

where λ is the unit constant failure rate. By integrating Equation (5.37) over the time interval [0, ∞], we get the following equation for the parallel-series network/system mean time to failure:

$$\begin{aligned} \mathrm{MTTF}_{ps} &= \int_0^{\infty} [1-(1-e^{-K\lambda t})^m]\, dt \\ &= \frac{1}{K\lambda}\sum_{i=1}^{m}\frac{1}{i} \end{aligned} \tag{5.38}$$

where MTTF_{ps} is the parallel-series network/system mean time to failure.

Example 5.5

Assume that in Example 5.4 the four units form a parallel-series system instead of series-parallel system. Calculate the parallel-series system mean time to failure, if each series subsystem contains two units.

Using the specified data in Equation (5.38) yield

$$\begin{aligned} \mathrm{MTTF}_{ps} &= \frac{1}{2(0.0004)}\left[1+\frac{1}{2}\right] \\ &= 1875\ \mathrm{h} \end{aligned}$$

Thus, the parallel-series system mean time to failure is 1875 h.

5.7 Standby System

In this case, one unit operates and $(m-1)$ units remain on standby as depicted by the block diagram in Figure 5.3. Each block in the diagram denotes a unit. When the operating unit fails, it is immediately replaced by one of the

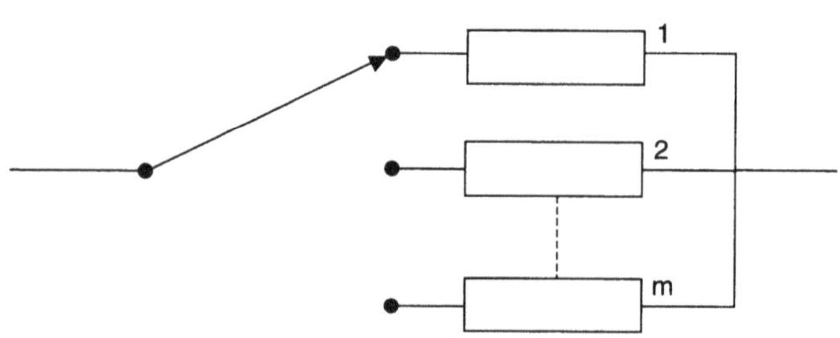

FIGURE 5.3
Standby system block diagram.

standbys. The system fails when all its units fail. The system reliability is given by [8]

$$R_{st}(t) = e^{-\lambda t} \sum_{i=0}^{m-1} (\lambda t)^i / i! \tag{5.39}$$

where

$R_{st}(t)$ = standby system reliability at time t
λ = unit failure rate
m = total number of units in the system (i.e., operating plus the standby unit)

The following assumptions are associated with Equation (5.39):

- The switching mechanism is perfect.
- The units are identical.
- The unit failure rate is constant.
- Units fail independently.
- Standby units remain as good as new.

By integrating Equation (5.39) over the time interval [0, ∞], we get the following equation for the standby system mean time to failure:

$$\begin{aligned} \text{MTTF}_{st} &= \int_0^\infty \left[e^{-\lambda t} \sum_{i=0}^{m-1} (\lambda t)^i / i! \right] dt \\ &= \frac{m}{\lambda} \end{aligned} \tag{5.40}$$

where MTTF_{st} is the standby system mean time to failure.

Example 5.6

A standby system is composed of two independent and identical units (i.e., one unit operating and the other on standby). The unit constant failure rate is 0.0007 failures per h. Calculate the following by using Equation (5.39) and Equation (5.40), respectively:

- System reliability for a 500-h mission.
- System mean time to failure.

By substituting the given data into Equation (5.39) and Equation (5.40), we get

$$\begin{aligned} R_{st}(500) &= e^{-(0.0007)(500)}[1 + (0.0007)(500)] \\ &= 0.9513 \end{aligned}$$

and

$$\text{MTTF}_{st} = \frac{2}{(0.0007)}$$
$$= 2857.14 \text{ h}$$

Thus, the standby system reliability and mean time to failure are 0.9513 and 2857.14 h, respectively.

5.8 Bridge Network

For time t dependent unit reliability, the five element bridge network reliability from Equation (4.21) is

$$\begin{aligned} R_b(t) = {} & 2R_1(t)R_2(t)R_3(t)R_4(t)R_5(t) + R_2(t)R_3(t)R_4(t) \\ & + R_1(t)R_3(t)R_5(t) + R_1(t)R_4(t) + R_2(t)R_5(t) \\ & - R_2(t)R_3(t)R_4(t)R_5(t) - R_1(t)R_2(t)R_3(t)R_4(t) \\ & - R_5(t)R_1(t)R_2(t)R_3(t) - R_1(t)R_3(t)R_4(t)R_5(t) \\ & - R_1(t)R_2(t)R_4(t)R_5(t) \end{aligned} \tag{5.41}$$

where R_b (t) is the bridge network reliability at time t and R_i (t) is the unit i reliability at time t, for $i = 1, 2, 3, 4, 5$.

For identical units Equation (5.41) becomes

$$R_b(t) = 2[R(t)]^5 - 5[R(t)]^4 + 2[R(t)]^3 + 2[R(t)]^2 \tag{5.42}$$

where R (t) is the unit reliability at time t.

For constant failure rate or exponentially distributed times to failure of a unit, by using Equation (5.2) in Equation (5.42) yields

$$R_b(t) = 2e^{-5\lambda t} - 5e^{-4\lambda t} + 2e^{-3\lambda t} + 2e^{-2\lambda t} \tag{5.43}$$

where λ is the unit constant failure rate.

By integrating Equation (5.43) over the time interval $[0, \infty]$, we get the following equation for the bridge network mean time to failure:

$$\begin{aligned} \text{MTTF}_b &= \int_0^{\infty} [2e^{-5\lambda t} - 5e^{-4\lambda t} + 2e^{-3\lambda t} + 2e^{-2\lambda t}]\, dt \\ &= \frac{49}{60\lambda} \end{aligned} \tag{5.44}$$

where MTTF_b is the bridge network mean time to failure.

Example 5.7

A bridge network is composed of five independent and identical units. The constant failure rate of a unit is 0.0006 failures/h. Calculate the network reliability for a 200-h mission and mean time to failure.

By substituting the specified data into Equation (5.43) yields

$$R_b(200) = 2e^{-5(0.0006)(200)} - 5e^{-4(0.0006)(200)} + 2e^{-3(0.0006)(200)} + 2e^{-2(0.0006)(200)}$$
$$= 0.9723$$

Similarly, using the given data value in Equation (5.44) yields

$$\text{MTTF}_b = \frac{49}{60(0.0006)}$$
$$= 1361.1 \text{ h}$$

Thus, the bridge network reliability and mean time to failure are 0.9723 and 1361.1 h, respectively.

5.9 Problems

1. Prove that the hazard rate of a series system is given by

$$\lambda_s(t) = \sum_{i=1}^{K} \lambda_i \tag{5.45}$$

where

$\lambda_s\,(t)$ = series system hazard rate
K = number of units in series
λ_i = unit i constant failure rate, for $i = 1, 2, 3, \dots, K$

2. Prove that the mean time to failure of a series system is given by

$$\text{MTTF}_s = \frac{1}{\Sigma_{i=1}^{K} \lambda_i} \tag{5.46}$$

where

MTTF_s = series system mean time to failure
K = number of units in series
λ_i = unit i constant failure rate, for $i = 1, 2, 3, \dots, k$

3. A computer has three independent and identical central processing units operating in parallel. The constant failure rate of the central

processing unit is 0.0001 failures/h. Calculate the computer mean time to failure with respect to central processing units.

4. Prove that series and parallel networks are the special cases of the k-out-of-m network.
5. An engineering system contains six active, independent, and identical units forming a series-parallel configuration. The constant failure rate of each unit is 0.0001 failures/h. Calculate the system mean time to failure, if each parallel subsystem contains two units.
6. Prove Equation (5.38).
7. A standby system is composed of three independent and identical units (i.e., one unit operating and the other two on standby). The unit constant failure rate is 0.0009 failures/h. Calculate the system reliability for a 200-h mission by using Equation (5.39).
8. A bridge network is composed of five independent and identical units. The constant failure rate of a unit is 0.0005 failures/h. Calculate the network reliability for a 300-h mission and mean time to failure.
9. Prove that a parallel network's reliability is given by

$$R_p(t) \approx 1 - \prod_{i=1}^{K} \lambda_i t^K \quad (5.47)$$

where

$R_p(t)$ = parallel network reliability at time t
K = total number of units in parallel
λ_i = constant failure rate of unit i, for $i = 1, 2, 3, \ldots, K$

10. Prove that a series network's reliability is given by

$$R_s(t) \approx 1 - K\lambda t \quad (5.48)$$

where

$R_s(t)$ = series network's reliability at time t
K = total number of units in series
λ = constant unit failure rate

References

1. Davis, D.J., An analysis of some failure data, *J. Am. Stat. Assoc.*, June 1952, pp. 113–150.
2. Shooman, M.L., *Probabilistic Reliability: An Engineering Approach*, McGraw-Hill, New York, 1968.

3. Weibull, W., A statistical distribution function of wide applicability, *J. Appl. Mech.*, Vol. 18, 1951, pp. 293–297.
4. Dhillon, B.S., A hazard rate model, *IEEE Trans. Reliability*, Vol. 29, 1979, p. 150.
5. Dhillon, B.S., *Design Reliability: Fundamentals and Applications*, CRC Press, Boca Raton, FL, 1999.
6. Drenick, R.F., The failure law of complex equipment, *J. Soc. Industrial Applied Mathematics*, Vol. 8, 1960, pp. 680–690.
7. MIL-HDBK-217, Reliability Prediction of Electronic Equipment, Department of Defence, Washington, D.C.
8. Sandler, G.H., *System Reliability Engineering*, Prentice Hall, Englewood Cliffs, NJ 1964.

6

Reliability Evaluation Methods

6.1 Introduction

Reliability evaluation is an important activity for ensuring the reliability of engineering products. It normally begins right from the conceptual design stage of products with specified reliability. Over the years, many reliability evaluation methods and techniques have been developed. Some examples of these methods and techniques are fault tree analysis (FTA), failure modes and effect analysis (FMEA), Markov method, network reduction method, and decomposition method. The use of these methods for a particular application depends on various factors including the specified requirement, the type of project under consideration, the specific need, and the inclination of the parties involved. For example, FMEA is often required in aerospace/defense related projects and FTA in nuclear power generation projects.

The ease of use and the requirement of specific experience of users (analysts) may vary from one method to another. For example, in the real world application the network reduction method is probably the easiest to use and it does not really require any specific experience from its users. In contrast, FMEA and FTA are relatively more demanding to perform and require considerable experience of analysts in the area of design.

This chapter presents a number of reliability evaluation methods.

6.2 Network Reduction Method

This is probably the simplest method for evaluating the reliability of systems composed of independent series and parallel subsystems. It sequentially reduces the parallel and series subsystems to equivalent hypothetical single units until the complete system itself becomes a single hypothetical unit. The bridge configurations or subsystems (if any) in the system can be converted to series and parallel equivalents by using delta-star conversions or the decomposition method [1, 2].

The main advantage of this approach is that it is easy to understand and apply. The method is demonstrated through the following example:

Example 6.1

A network representing an engineering system with independent units is shown in Figure 6.1(a). Each block in the figure denotes a unit. The reliability R_j of unit j, for $j = 1, 2, 3, \ldots, 7$ is given. Determine the network reliability by using the network reduction method.

First we have identified subsystems A, B, and C of the network as shown in Figure 6.1(i). The subsystem A has three units in series; thus we reduce them to a single hypothetical unit as follows:

$$\begin{aligned} R_A &= R_1 R_2 R_3 \\ &= (0.5)(0.6)(0.7) \\ &= 0.21 \end{aligned}$$

where R_A is the reliability of subsystem A.

Thus, subsystem A has been reduced to a single hypothetical unit having reliability 0.21. The reduced network is shown in Figure 6.1(ii). Now, this network is made up of two parallel subsystems B and C acting in series. Thus, we reduce subsystem B to a single hypothetical unit as follows:

$$\begin{aligned} R_B &= 1-(1-R_5)(1-R_6)(1-R_7) \\ R_B &= 1-(1-0.7)(1-0.8)(1-0.9) \\ &= 0.994 \end{aligned}$$

where R_B is the reliability of subsystem B.

Thus, subsystem B has been reduced to a single hypothetical unit having reliability 0.994. The reduced network is shown in Figure 6.1(iii). This network contains subsystem C and a hypothetical unit, representing subsystem B, in series. In similar manner to subsystem B, we reduce subsystem C to a single hypothetical unit:

$$\begin{aligned} R_C &= 1-(1-R_A)(1-R_4) \\ &= 1-(1-0.21)(1-0.8) \\ &= 0.842 \end{aligned}$$

where R_C is the subsystem C reliability.

Similarly, the reduced network is shown in Figure 6.1(iv). This network is composed of two hypothetical units, representing subsystems B and C, in series.

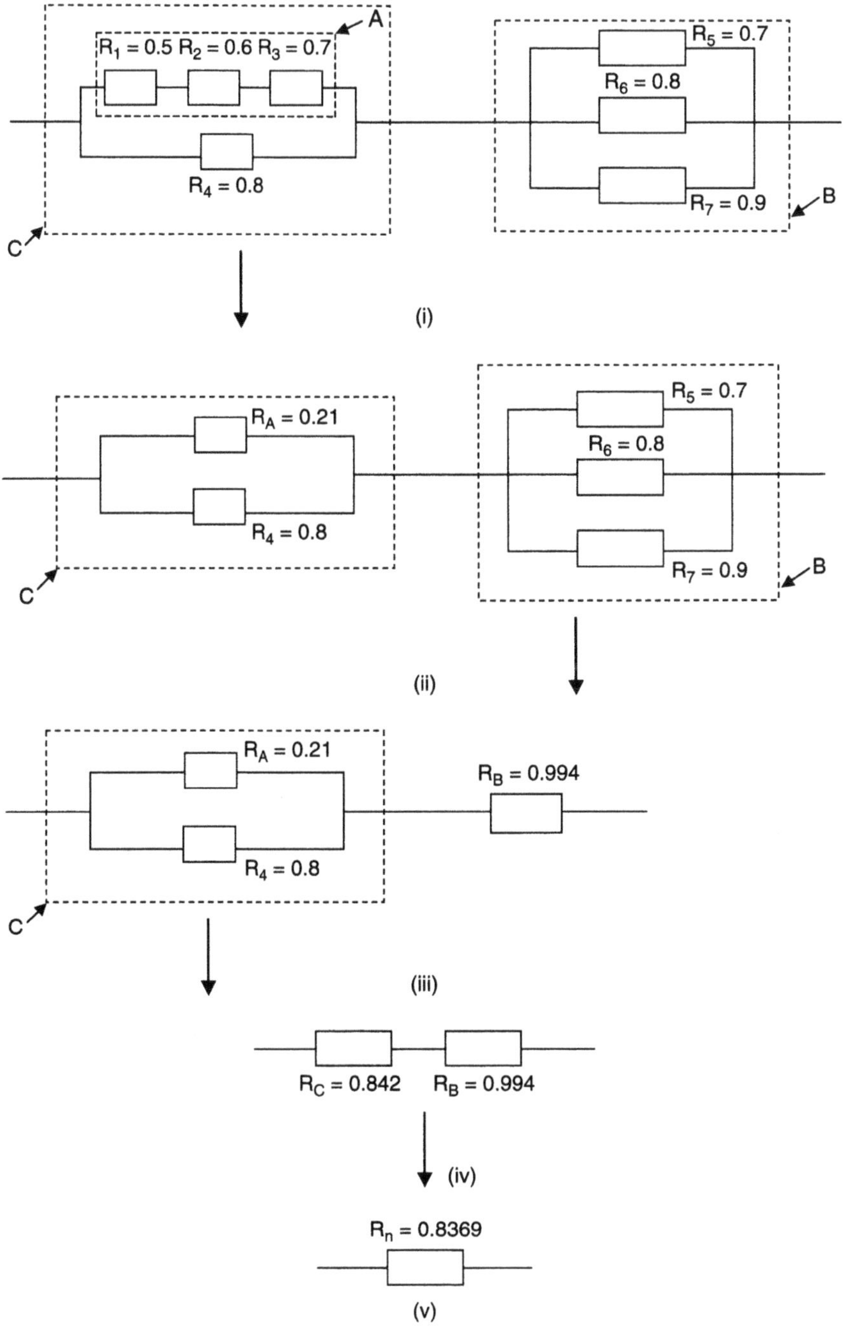

FIGURE 6.1
Diagrammatic steps of the network reduction method: (i) original network; (ii) reduced network; (iii) reduced network; (iv) reduced network; (v) single hypothetical unit.

The reliability of this network is given by

$$R_n = R_B R_C$$
$$= (0.994)(0.842)$$
$$= 0.8369$$

where R_n is the reliability of the whole network shown in Figure 6.1(i).

All in all, by using the network reduction method, the Figure 6.1(i) network was reduced to a single hypothetical unit having reliability 0.8369 (Figure 6.1(v)); which is the whole network's reliability.

6.3 Fault Tree Analysis (FTA) Method

This is a widely used method in industry to evaluate reliability of engineering systems. It was developed in the early 1960s at the Bell Telephone Laboratories to analyze Minuteman Launch Control System with respect to reliability and safety [3].

FTA is concerned with fault events and a fault tree may simply be described as a logical representation of the relationship of primary or basic fault events that lead to the occurrence of a specified undesirable fault event known as the "top event." Furthermore, a fault tree is depicted using a tree structure with logic gates such as AND and OR.

6.3.1 Fault Tree Symbols and Basic Steps for Developing a Fault Tree

The FTA approach makes use of a large number of symbols [4, 5]. Five commonly used symbols in fault tree construction are shown in Figure 6.2.

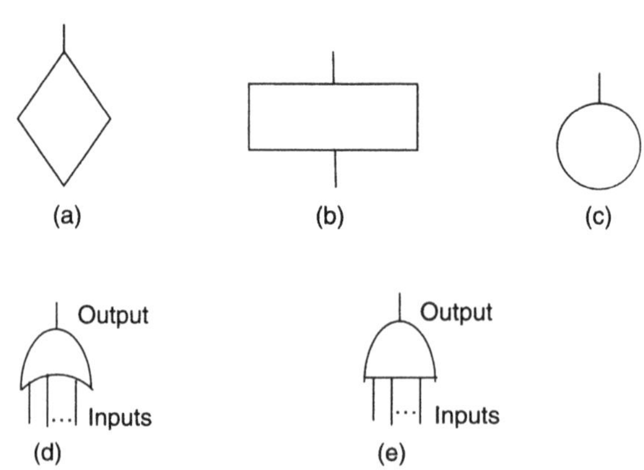

FIGURE 6.2
Common fault tree symbols: (a) diamond, (b) rectangle, (c) circle, (d) OR gate, (e) AND gate.

Each of these symbols is described below.

- **Diamond:** This denotes a fault event whose causes have not been fully developed either due to lack of information or due to lack of interest.
- **Rectangle:** This denotes a fault event that results from the combination of fault events through the input of a logic gate.
- **Circle:** This represents a basic fault event or the failure of an elementary component. The event's occurrence probability and failure and repair rates are normally obtained from empirical data.
- **OR gate:** This denotes that an output fault occurs if any one or more of the input fault events occur.
- **AND gate:** This denotes that an output fault event occurs if and only if all the input fault events occur.

The following four basic steps are generally followed in developing a fault tree [2]:

- Define the undesired event (i.e., the top event) of the system under consideration, to be studied.
- Understand thoroughly the system and its intended application.
- Determine the higher-order functional fault events in order to obtain the predefined system fault condition cause. Furthermore, continue FTA for determining the logical interrelationship of lower level fault events that can cause them.
- Construct a fault tree of logical relationships among input fault events.
- Evaluate the fault tree (i.e., qualitatively/quantitatively).

Example 6.2

A windowless room has two light bulbs and one switch. The room can only be dark if the switch fails to close, there is no electricity, or both the bulbs burn out. Develop a fault tree for the occurrence of undesired event (top event) "dark room," more specifically, the room without light.

A fault tree for the example is shown in Figure 6.3. Each fault event in the figure is labeled as $E_1, E_2, E_3, \ldots, E_8$.

6.3.2 Probability Evaluation of Fault Trees

When the probability of occurrence of basic fault events is known, the probability of occurrence of the top event can be calculated. This requires first estimating the probability of occurrence of output fault events of lower and intermediate logic gates such as AND and OR. The output fault event

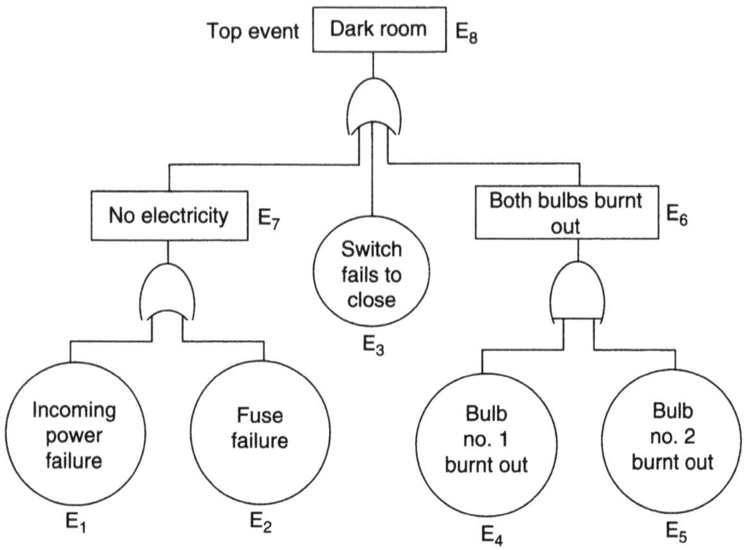

FIGURE 6.3
Fault tree for the occurrence of top event: dark room.

occurrence probability of AND and OR gates is obtained, respectively, as follows [2, 4, 5]:

1. **AND gate:** The probability of occurrence of the AND gate output fault event E_o is given by

$$P(E_o) = \prod_{i=0}^{n} P(E_i) \tag{6.1}$$

where
$P(E_o)$ = occurrence probability of the AND gate output fault event, E_o
n = total number of independent input fault events
$P(E_i)$ = probability of occurrence of input fault event E_i, for $i = 1, 2, 3, \ldots, n$

2. **OR gate:** The probability of occurrence of the OR gate output fault event X_o is given by

$$P(X_o) = 1 - \prod_{i=1}^{k} \{1 - P(X_i)\} \tag{6.2}$$

where
$P(X_o)$ = occurrence probability of the OR gate output fault event, X_o
k = total number of independent input fault events
$P(X_i)$ = probability of occurrence of input fault event X_i, for $i = 1, 2, 3, \ldots, k$

Example 6.3

Assume that in Figure 6.3, the probability of occurrence of basic events E_1, E_2, E_3, E_4, and E_5 are 0.05, 0.03, 0.06, 0.1, and 0.1, respectively. Calculate the probability of occurrence of the top event: dark room and then show Figure 6.3 fault tree with given and calculated fault event occurrence probability values.

Using the specified data in Equation (6.1), the occurrence probability of event E_6 (i.e., both bulbs burnt out) is

$$\begin{aligned} P(E_6) &= P(E_4)P(E_5) \\ &= (0.1)(0.1) \\ &= 0.01 \end{aligned}$$

Similarly, by inserting the given data into Equation (6.2), the probability of occurrence of event E_7 (i.e., no electricity) is

$$\begin{aligned} P(E_7) &= 1 - \{1 - P(E_1)\}\{1 - P(E_2)\} \\ &= 1 - \{1 - 0.05\}\{1 - 0.03\} \\ &= 0.0785 \end{aligned}$$

Using the above calculated values and the given data in Equation (6.2) yields

$$\begin{aligned} P(E_8) &= 1 - \{1 - P(E_3)\}\{1 - P(E_7)\}\{1 - P(E_6)\} \\ &= 1 - \{1 - 0.06\}\{1 - 0.0785\}\{1 - 0.01\} \\ &= 0.1424 \end{aligned}$$

where $P(E_8)$ is the probability of occurrence of fault event E_8 (i.e., dark room).

Thus, the probability of having no light in the room is 0.1424 and the Figure 6.3 fault tree with given and calculated fault event occurrence probability values is shown in Figure 6.4

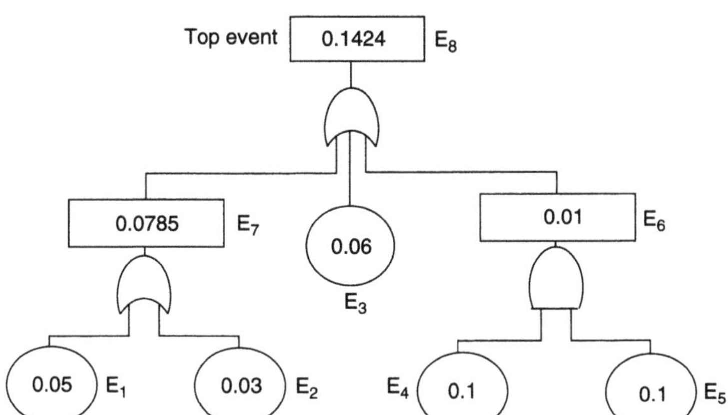

FIGURE 6.4
Fault tree of Figure 6.3 with given and calculated fault event occurrence probability values.

6.4 Markov Method

This is a widely used method in industry to perform various types of reliability analysis. The method is named after a Russian mathematician, Andrei Andreyevich Markov (1856–1922). Markov method is quite useful to model systems with dependent failure and repair modes and is based on the following assumptions [6]:

- The probability of transition from one system state to another in the finite time interval Δt is given by $\alpha \Delta t$, where α is the transition rate (e.g., constant failure or repair rate of an item) from one system state to another.
- The probability of more than one transition in time interval Δt from one state to the next state is negligible (e.g., $(\alpha \Delta t)(\alpha \Delta t) \rightarrow 0$).
- The occurrences are independent of each other.

The application of this method is demonstrated through the following example:

Example 6.4

An engineering system can either be in a working state or a failed state. The system state space diagram is shown in Figure 6.5. The numerals in boxes denote the system state. The system fails at a constant failure rate λ.

Develop expressions for system reliability, unreliability, and mean time to failure.

With the aid of Markov method, we write down the following equations for the Figure 6.5 diagram:

$$P_0(t+\Delta t) = P_0(t)(1-\lambda \Delta t) \tag{6.3}$$

$$P_1(t+\Delta t) = P_1(t) + (\lambda \Delta t)P_0(t) \tag{6.4}$$

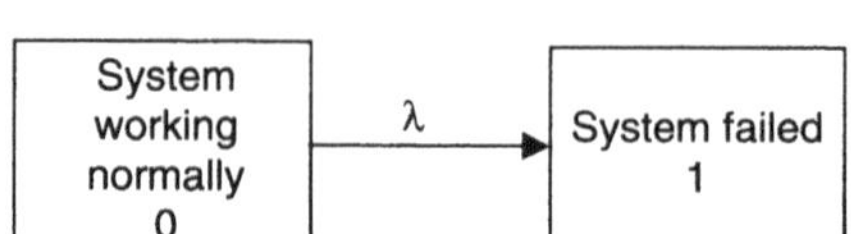

FIGURE 6.5
System state space diagram.

where

$P_i\,(t + \Delta t)$ = probability that at time $t + \Delta t$ the system is in state i, $i = 0$ (working normally), $i = 1$ (failed)

$P_i\,(t)$ = probability that at time t the system is in state i, $i = 0$ (working normally), $i = 1$ (failed)

λ = system constant failure rate

$\lambda\Delta t$ = probability of system failure in finite time interval Δt

$(1 - \lambda\Delta t)$ = probability of no failure in time interval Δt when the system is in state 0

In the limiting case, Equation (6.3) and Equation (6.4) become

$$\underset{\Delta t \to 0}{\text{limit}} \frac{P_0(t+\Delta t) - P_0(t)}{\Delta t} = \frac{dP_0(t)}{dt} = -\lambda P_0(t) \tag{6.5}$$

and

$$\underset{\Delta t \to 0}{\text{limit}} \frac{P_1(t+\Delta t) - P_1(t)}{\Delta t} = \frac{dP_1(t)}{dt} = \lambda P_0(t) \tag{6.6}$$

At time $t = 0$, $P_0\,(0) = 1$ and $P_1\,(0) = 0$.

Solving Equation (6.5) and Equation (6.6), we get

$$P_0(t) = R_S(t) = e^{-\lambda t} \tag{6.7}$$

$$P_1(t) = UR_S(t) = (1 - e^{-\lambda t}) \tag{6.8}$$

where $R_S\,(t)$ is the system reliability at time t and $UR_S\,(t)$ is the system unreliability at time t. The system mean time to failure is given by [7]

$$\begin{aligned} \text{MTTF}_S &= \int_0^\infty R_S(t)dt = \int_0^\infty e^{-\lambda t}dt \\ &= \frac{1}{\lambda} \end{aligned} \tag{6.9}$$

where MTTF_S is the system mean time to failure. Thus, expressions for system reliability, unreliability, and mean time to failure are given by Equation (6.7), Equation (6.8), and Equation (6.9), respectively.

6.5 Decomposition Method

This method is used to evaluate reliability of complex systems. It decomposes complex systems into simpler subsystems by applying the conditional probability theory. At the end, system reliability is obtained by combining the reliability measures of subsystems.

The method begins by first selecting the key element or unit to be used to decompose a given network/system. The poor choice of this key element leads to poor efficiency of computing system reliability. Nonetheless, the past experience usually plays an instrumental role in selecting the right key element.

First, the method assumes that the key element/unit, say x, is replaced by another element that never fails (i.e., 100% reliable) and then it assumes that the key element is 100% unreliable (i.e., it is completely removed from the system or network). Under this scenario, the overall system/network reliability is given by [6]

$$R_S = P(x)P(\text{system good}/x \text{ good}) + P(\overline{x})P(\text{system good}/x \text{ fails}) \qquad (6.10)$$

where

R_S = system reliability
P (system good/x good) = reliability of the system when x is 100% reliable
P (system good/x fails) = reliability of the system when x is 100% unreliable
P (x) = reliability of the key element x
P ($\overline{x}$) = unreliability of the key element x

Similarly, the overall system/network unreliability is expressed by [6]

$$UR_S = P(x)P(\text{system fails}/x \text{ good}) + P(\overline{x})P(\text{system fails}/x \text{ fails}) \qquad (6.11)$$

where

UR_S = system unreliability
P (system fails/x good) = unreliability of the system when x is 100% reliable
P (system fails/x fails) = unreliability of the system when x is 100% unreliable

The application of this method is demonstrated through the following example.

Example 6.5

A five independent unit bridge network is shown in Figure 6.6. Each block in the diagram denotes a unit and each unit's reliability is denoted by R_i, for

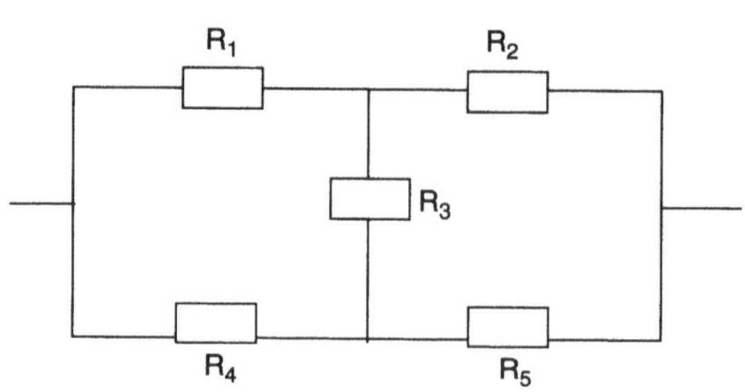

FIGURE 6.6
A five dissimilar unit bridge network.

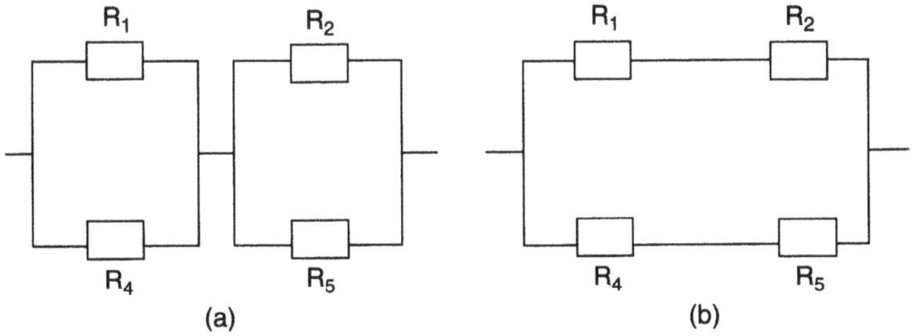

FIGURE 6.7
Reduced networks of Figure 6.6 diagram: (a) For a 100% reliable key element, (b) For a 100% unreliable key element.

$i = 1, 2, 3, \ldots, 5$. Develop an expression for the network reliability by utilizing the decomposition approach.

First of all, in this example we identify the Figure 6.6 unit with reliability R_3 as our key element, say x. Thus, by replacing the key element in Figure 6.6 with a 100% reliable unit and then with a 100% unreliable unit results in Figure 6.7(a) and Figure 6.7(b) diagrams, respectively.

Using the network reduction method, we obtain the following reliability expression for Figure 6.7(a):

$$R_{Sp} = [1-(1-R_1)(1-R_4)][1-(1-R_2)(1-R_5)] \tag{6.12}$$

where R_{Sp} is the series-parallel network reliability (i.e., the system reliability when the key element is 100% reliable).

For identical units (i.e., $R_1 = R_2 = R_4 = R_5 = R$) Equation (6.12) becomes

$$\begin{aligned} R_{Sp} &= [1-(1-R)^2]^2 \\ &= (2R-R^2)^2 \end{aligned} \tag{6.13}$$

where R is the unit reliability. Similarly, by utilizing the network reduction approach, we get the following reliability expression for Figure 6.7 (b):

$$R_{pS} = 1-(1-R_1R_2)(1-R_4R_5) \tag{6.14}$$

where R_{pS} is the parallel-series network reliability (i.e., the system reliability when the key element is 100% unreliable).

For identical units, Equation (6.14) becomes

$$\begin{aligned} R_{pS} &= 1-(1-R)^2 \\ &= 2R^2 - R^4 \end{aligned} \tag{6.15}$$

The reliability and unreliability of the key element x, respectively, are given by

$$P(x) = R_3 \tag{6.16}$$

and

$$P(\bar{x}) = 1 - R_3 \tag{6.17}$$

For $R_3 = R$, Equation (6.16) and Equation (6.17) become

$$P(x) = R \tag{6.18}$$

and

$$P(\bar{x}) = (1 - R) \tag{6.19}$$

Substituting Equation (6.12), Equation (6.14), Equation (6.16), and Equation (6.17) into Equation (6.10) yields

$$R_S = R_3[1-(1-R_1)(1-R_4)][1-(1-R_2)(1-R_5)] + (1-R_3)[1-(1-R_1R_2)(1-R_4R_5)] \tag{6.20}$$

For identical units, inserting Equation (6.13), Equation (6.15), Equation (6.18), and Equation (6.19) into Equation (6.10), we get

$$\begin{aligned} R_S &= R(2R - R^2)^2 + (1-R)(2R^2 - R^4) \\ &= 2R^2 + 2R^3 - 5R^4 + 2R^5 \end{aligned} \tag{6.21}$$

Thus, Equation (6.20) and Equation (6.21) are the reliability expressions for Figure 6.6 network with nonidentical and identical units, respectively.

6.6 Failure Modes and Effect Analysis (FMEA) Method

This is a widely used method in industry to analyze engineering systems with respect to reliability. The history of the FMEA may be traced back to the early 1950s with the development of flight control systems when the U.S. Navy's bureau of aeronautics developed a reliability-related requirement called "Failure Analysis" [8]. Subsequently, "Failure Analysis" became known as "Failure Effect Analysis" and then "Failure Modes and Effect Analysis" [9].

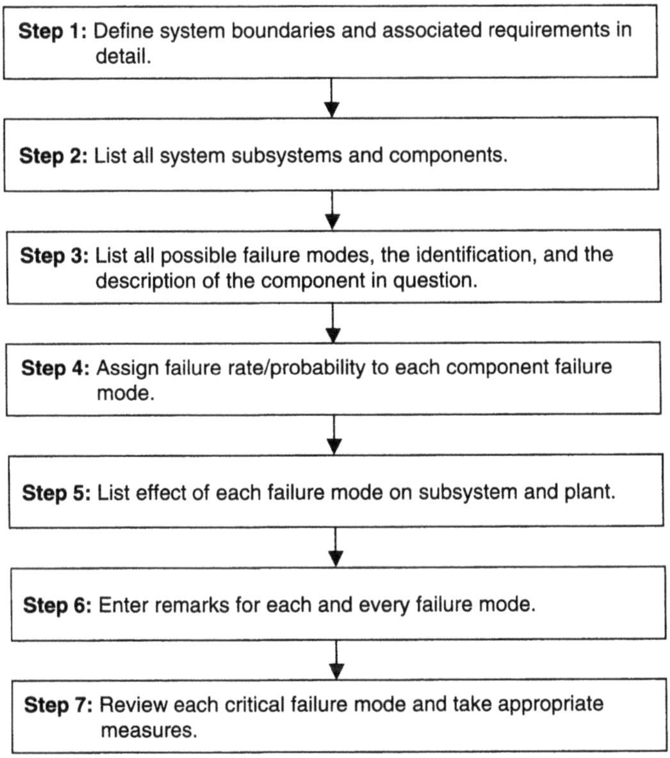

FIGURE 6.8
Steps for performing FMEA.

Needless to say, to assure the desired reliability of space systems National Aeronautics and Space Administration (NASA) extended FMEA to classify each potential failure effect according to its severity and called it failure mode effects and criticality analysis (FMECA) [10].

A comprehensive list of publications on FMEA/FMECA is given in Ref. [11]. FMEA may simply be described as an approach used to perform analysis of each potential failure mode in the system to examine the results or effects of these failure modes on the system [12]. Seven main steps used to perform FMEA are given in Figure 6.8 [2].

FMEA is described in detail in Ref. [13].

6.7 Common Cause Failure Analysis Method

There are many instances in industry when multiple units have failed due to a single cause [14]. A common-cause failure may simply be described as any instance where multiple units fail due to a single cause. Some of the main causes for the occurrence of common-cause failures in engineering systems are as follows [15]:

- Poor design
- Common operational environment (e.g., humidity, vibration, dust, temperature, and moisture)
- External catastrophic events (e.g., flood, earthquake, fire, and tornado)
- Operations and maintenance errors
- Functional deficiencies
- Common manufacturer of parts/items
- Common external power source

Needless to say, common-cause failures are important factors in reliability analysis of engineering systems. Their occurrence must be taken into account in reliability prediction, otherwise the predicted reliability will not represent the real picture.

Over the years, various methods and models have been developed to take into consideration the occurrence of common-cause failures in reliability evaluation of engineering systems [2]. Here, we present one such method. The method is based on the assumption that all common-cause failures associated with a redundant system can be represented by a single hypothetical unit. In turn, this unit is placed in series with the redundant system.

Some examples of this redundant system are a parallel network, a k-out-of-n unit system, and a bridge configuration. Nonetheless, here we demonstrate this method for the redundant system being a parallel network as shown in Figure 6.9. Each block in the figure denotes a unit and the symbols n and cf denote the number of units in parallel and the hypothetical unit representing all common-cause failures associated with the parallel network, respectively.

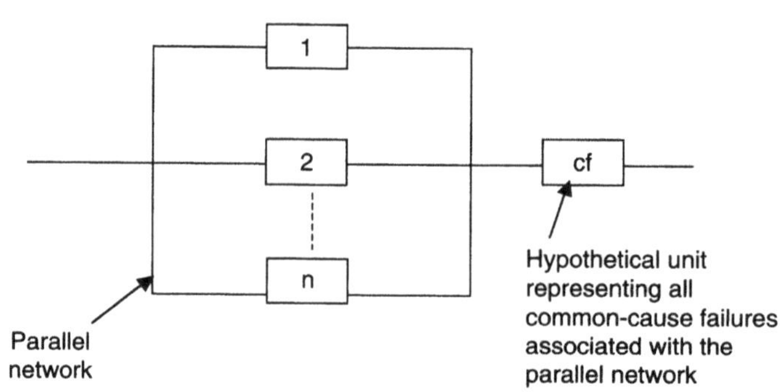

FIGURE 6.9
Block diagram of a parallel system with common-cause failures.

The reliability of the Figure 6.9 diagram is

$$R_{pc} = [1-(1-R)^n]\, R_{cf} \tag{6.22}$$

where

R_{pc} = reliability of the parallel system with common-cause failures
N = total number of independent and identical units in parallel
R = unit reliability
R_{cf} = reliability of the hypothetical unit representing all common-cause failures associated with the parallel network

For exponentially distributed unit and common-cause failure times, the total failure rate of a unit is expressed as follows [16, 17]:

$$\lambda_T = \lambda + \lambda_{cf} \tag{6.23}$$

where

λ_T = unit total failure rate
λ = unit independent mode (i.e., noncommon cause) failure rate
λ_{cf} = hypothetical unit (i.e., the unit which represents all common-cause failures associated with the parallel network) failure rate

Hence, we define a fraction of unit failures that are common-cause type as follows:

$$\gamma = \frac{\lambda_{cf}}{\lambda_T} \tag{6.24}$$

Therefore,

$$\lambda_{cf} = \gamma \lambda_T \tag{6.25}$$

Substituting Equation (6.25) into Equation (6.23) and rearranging, we get

$$\lambda = (1-\gamma)\lambda_T \tag{6.26}$$

A general equation for a unit's reliability is [7]

$$R(t) = e^{-\int_0^t \lambda(x)dx} \tag{6.27}$$

where $R(t)$ is the unit reliability at time t and $\lambda(t)$ is the unit hazard rate (i.e., time dependent failure rate).

Thus, using Equation (6.25) and Equation (6.26), in Equation (6.27), respectively, we get

$$R_{cf}(t) = e^{-\gamma \lambda_T t} \tag{6.28}$$

and

$$R(t) = e^{-(1-\gamma)\lambda_T t} \tag{6.29}$$

where $R_{cf}(t)$ is the hypothetical unit, representing common-cause failures, reliability at time t and $R(t)$ is the unit reliability.

Substituting Equation (6.28) and Equation (6.29) into Equation (6.22) yields

$$R_{pc}(t) = [1-(1-e^{-(1-\gamma)\lambda_T t})^n]\, e^{-\gamma \lambda_T t} \tag{6.30}$$

where R_{pc} (t) is the reliability of the parallel system with common-cause failures at time t. The system mean time to failure is given by [7]

$$\begin{aligned} \mathrm{MTTF}_{pc} &= \int_0^\infty R_{pc}(t)\,dt \\ &= \sum_{i=1}^{n} (-1)^{i+1} \binom{n}{i} \Big/ \lambda \{i-(i-1)\gamma\} \end{aligned} \tag{6.31}$$

where MTTF_{pc} is the mean time to failure of the parallel system with common-cause failures.

$$\binom{n}{i} = \frac{n!}{i!(n-i)!} \tag{6.32}$$

Example 6.6

An engineering system is composed of three independent and identical units in parallel. The system is subjected to common-cause failures with their constant occurrence rate of 0.0002 failures/h. The unit constant failure rate (i.e., noncommon-cause failure rate) is 0.004 failures/h. Calculate the system reliability for a 500-h mission with and without the occurrence of common-cause failures.

Using the above specified data values in Equation (6.23) and Equation (6.24), we get

$$\begin{aligned} \lambda_T &= 0.004 + 0.0002 \\ &= 0.0042 \text{ failures/h} \end{aligned}$$

and

$$\gamma = \frac{0.0002}{0.0042}$$
$$= 0.0476$$

Substituting the above two calculated values and the specified data into Equation (6.30) yields

$$R_{pc}(500) = [1-(1-e^{-(1-0.0476)(0.0042)(500)})^3]e^{-(0.0476)(0.0042)(500)}$$
$$= 0.3199$$

When there are no common-cause failures, $\lambda_{cf} = 0$ and $\gamma = 0$. Thus, using Equation (6.24) and Equation (6.30), we get

$$R_{pc}(t) = 1-(1-e^{-\lambda t})^n \tag{6.33}$$

Inserting the given data values into Equation (6.33) yields

$$R_{pc}(500) = 1-(1-e^{-(0.004)(500)})^3$$
$$= 0.3535$$

Thus, the system reliability with and without the occurrence of common-cause failures is 0.3199 and 0.3535, respectively. Obviously, the occurrence of common-cause failures results in lower system reliability.

6.8 Problems

1. Define a common-cause failure.
2. What are the main causes for the occurrence of common-cause failures?
3. Describe the failure modes and effect analysis (FMEA) method.
4. Draw the symbols used to denote AND and OR logic gates. Discuss the functions of these two gates.
5. Assume that a windowless room has four light bulbs and one switch. The room can only be dark if the switch fails to close, there is no electricity, or all the bulbs burn out. Develop a fault tree for the occurrence of undesired event (top event) "dark room."
6. Prove using Equation (6.5) and Equation (6.6) that a system's reliability and unreliability are given by Equation (6.7) and Equation (6.8), respectively.

7. Discuss the following reliability evaluation methods:
   - Markov method
   - Decomposition method
8. An engineering system is composed of two independent and identical units in parallel. The system is subjected to common-cause failures with their constant occurrence rate of 0.0001 failures/h. The unit constant failure rate (i.e., noncommon-cause failure rate) is 0.006 failures/h. Calculate the system reliability for a 100-h mission with and without the occurrence of common-cause failures. Comment on the end results.
9. What are the advantages of the network reduction method?
10. Compare the fault tree analysis (FTA) method with the failure modes and effect analysis (FMEA) method.

References

1. Dhillon, B.S., The analysis of the reliability of multistate device networks, Ph.D. dissertation, 1975, Available from the National Library of Canada, Ottawa, Canada.
2. Dhillon, B.S., Singh, C., *Engineering Reliability: New Techniques and Applications*, John Wiley & Sons, New York, 1981.
3. Haasl, D.F., Advanced concepts in fault tree analysis, System Safety Symposium, 1965, Available from the University of Washington Library, Seattle.
4. *Fault Tree Handbook*, Report No. NUREG-0492, U.S. Nuclear Regulatory Commission, Washington, D.C., 1981.
5. Risk Analysis Using the Fault Tree Technique, Flow Research Report, Flow Research, Washington, D.C., 1973.
6. Shooman, M.L., *Probabilistic Reliability: An Engineering Approach*, McGraw-Hill, New York, 1968.
7. Dhillon, B.S., *Mechanical Reliability: Theory, Models, and Applications*, American Institute of Aeronautics and Astronautics, Washington, D.C., 1988.
8. MIL-F-18372 (Aer), *General Specification for Design, Installation, and Test of Aircraft Flight Control Systems*, Bureau of Naval Weapons, Department of the Navy, Washington, D.C., Para. 3.5.2.3.
9. Coutinho, J.S., *Failure Effect Analysis*, Trans. of the New York Academy of Sciences, Vol. 26, Series II, 1963–1964, pp. 564–584.
10. Jordan, W.E., Failure modes, effects, and criticality analyses, *Proceedings of the Annual Reliability and Maintainability Symposium*, 1972, pp. 30–37.
11. Dhillon, B.S., Failure modes and effects analysis: bibliography, *Microelectronics and Reliability*, Vol. 32, 1992, pp. 719–731.
12. Omdahl, T.P., Ed., *Reliability, Availability, and Maintainability (RAM) Dictionary*, American Society for Quality Control (ASQC) Press, Milwaukee, WI, 1988.
13. Palady, P., *Failure Modes and Effect Analysis*, PT Publications, West Palm Beach, FL, 1995.

14. Gangloff, W.C., Common mode failure analysis, *IEEE Transactions on Power Apparatus Systems*, Vol. 94, February 1975, pp. 27–30.
15. WASH (NUREG—75/014), Reactor Safety Study, U.S. Nuclear Regulatory Commission, Washington, D.C., October 1975.
16. Fleming K.N., A redundant model for common mode failures in redundant safety systems, *Proceedings of the Sixth Pittsburgh Annual Modeling and Simulation Conference*, 1975, pp. 579–581.
17. Dhillon, B.S. and Proctor, C.L., Common mode failure analysis of reliability networks, *Proceedings of the Annual Reliability and Maintainability Symposium*, 1977, pp. 404–408.

7

Reliability Testing

7.1 Introduction

Testing is an important element of any engineering product development program. It may simply be described as subjecting the product under consideration to conditions that identify its weaknesses, modes of failures, and behavior characteristics. Testing is performed in two distinct modes (i.e., standard or accelerated). In the case of standard mode, the tests are conducted at typical operating parameters and at ambient temperature. Moreover, the actual operation time is considered as the test time. In the case of accelerated mode, parameters such as voltage, temperature, or frequency of cycling are varied above their normal values to reduce test time, or the test could simply be a test such as sudden death testing.

Reliability testing is an important component of testing and it is one of the most important reliability activities of a reliability program. The main objective of the reliability testing is to obtain as much information as possible regarding item/product failures, in particular, the item's tendency to fail and the consequences of failures. Nonetheless, it may be said that a good reliability testing program requires minimal testing and provides maximum information concerning failures. Reference [1] provides a comprehensive list of publications on reliability testing and two important publications on the topic are listed as Refs. [2, 3].

This chapter presents various important aspects of reliability testing.

7.2 Types of Reliability Testing

Reliability testing may be classified under three distinct categories as shown in Figure 7.1 [4]. These are reliability development and demonstration testing, qualification and acceptance testing, and operational testing.

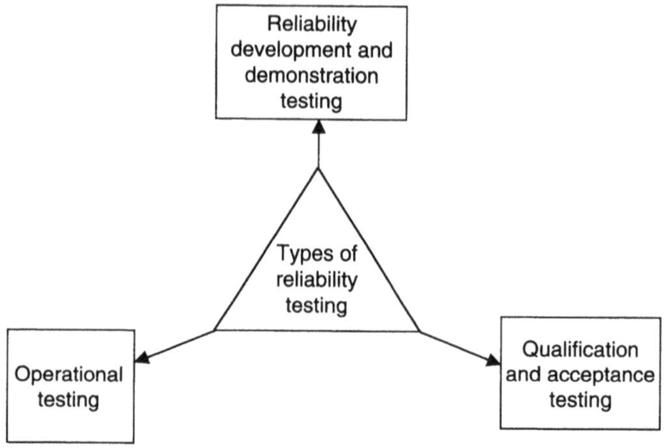

FIGURE 7.1
Types of reliability testing.

Reliability development and demonstration testing is concerned with satisfying various objectives including to determine if any design changes are required, to verify the degree of improvements in design reliability, and to determine if the design is to be improved to meet the reliability requirements. The nature of this type of testing depends on factors such as the type of system/subsystem under consideration, and the level of complexity involved. In order to satisfy reliability development and demonstration testing objectives in an effective manner, the generated test data must be of a type that permits insight into the failure probabilities/failure effects for a given design.

Qualification and acceptance testing has the following two fundamental objectives:

- To determine if a particular design is qualified for its intended use.
- To decide if a part/assembly/end item is to be accepted or rejected.

It is to be noted that these two objectives differ from the objectives of other reliability tests, particularly with respect to accept/reject criteria. In regard to the parts and materials to be used in the product or item under development, the qualification and acceptance testing begins early in the program. More specifically, this type of testing involves the usual testing and screening the quality-control function of incoming parts and materials.

Operational testing is concerned with meeting objectives which are as follows [4, 5]:

- To verify the results of reliability analysis performed during the product/item design and development.
- To provide useful data for subsequent activities or projects.

- To provide data indicating desirable changes to operating policies and procedures in regard to reliability/maintainability.

7.3 Tests for Determining the Validity of an Item's Exponentially Distributed Times to Failure

Exponential distribution is probably the most widely used statistical distribution in reliability studies to represent failure times of engineering items. Moreover, past experiences indicate that the failure times of many engineering items, in fact, do follow the exponential distribution, in particular, the electronic ones. Over the years, many mathematically-based tests have been developed to verify the assumption that a given set of failure time data belong to the exponential distribution [6, 9]. This section presents two of these tests.

7.3.1 Test I

This test is known as the Bartlett test and its statistic is defined as [7, 10].

$$S_{bm} = 12m^2\left[\ln M - \frac{N}{m}\right] \Big/ (6m + m + 1) \tag{7.1}$$

where

$$N \equiv \sum_{j=1}^{m} lnt_j \tag{7.2}$$

$$M \equiv \frac{1}{m}\sum_{j=1}^{m} t_j \tag{7.3}$$

m = total number of times to failure in the sample
t_j = jth time to failure

A sample of minimum 20 failure times is required for the test to discriminate effectively. Nonetheless, if the times to failure are exponentially distributed, the S_{bm} is distributed as chi-square with $(m - 1)$ degrees of freedom. Thus, a two-tailed chi-square criterion is used [9]. The following example demonstrates the application of this method:

TABLE 7.1

An Engineering System's Failure Times (in Days)

5	20	55	85	180
10	40	62	95	200
15	30	70	100	220
20	45	80	140	250
25	50	75	150	275

Example 7.1

An engineering system was closely monitored with respect to failures over a period of time. A sample of its 25 failure times (in days) is given in Table 7.1. Determine by using the Bartlett test at 95% confidence level that Table 7.1 failure times belong to an exponential distribution.

Using the Table 7.1 data in Equation (7.2) yields

$$\begin{aligned} N &= \ln 5 + \ln 10 + \ln 15 + \ln 20 + \ln 25 + \cdots + \ln 275 \\ &= 102.29 \end{aligned}$$

Similarly, by substituting the Table 7.1 data into Equation (7.3), we get

$$\begin{aligned} M &= \frac{1}{25}[5 + 10 + 15 + 20 + 25 + 20 + 40 + \cdots + 275] \\ &\equiv \frac{2297}{25} \\ &= 91.88 \end{aligned}$$

Using the above two results and the given data in Equation (7.1) yields

$$\begin{aligned} S_{b25} &= 12(25)^2 \left[\ln 91.88 - \frac{102.29}{25}\right] \Big/ \{6(25) + 25 + 1\} \\ &= 18.28 \end{aligned}$$

From Table 7.2 [10, 11] for a two-tailed test with 95% confidence level, the critical values of

$$\begin{aligned} \chi^2\left[\frac{\alpha}{2}, (m-1)\right] &= \chi^2\left[\frac{0.05}{2}, (25-1)\right] \\ &= \chi^2[0.025, 24] \\ &= 39.36 \end{aligned}$$

TABLE 7.2
Chi-Square Distribution Values

Degrees of Freedom	Probability				
	0.975	0.95	0.90	0.05	0.025
1	0.001	0.004	0.016	3.84	5.02
2	0.05	0.1	0.21	5.99	7.38
3	0.22	0.35	0.58	7.81	9.35
4	0.48	0.71	1.06	9.49	11.14
5	0.83	1.15	1.61	11.07	12.83
6	1.24	1.64	2.20	12.59	14.45
7	1.69	2.17	2.83	14.07	16.01
8	2.18	2.73	3.49	15.51	17.53
9	2.7	3.33	4.17	16.92	19.02
10	3.25	3.94	4.86	18.31	20.48
11	3.82	4.58	5.58	19.67	21.92
12	4.4	5.23	6.30	21.02	23.34
13	5.01	5.89	7.04	22.36	24.74
14	5.63	6.57	7.79	23.68	26.12
15	6.26	7.26	8.55	24.99	27.49
16	6.91	7.96	9.31	26.29	28.84
17	7.56	8.67	10.08	27.59	30.19
18	8.23	9.39	10.86	28.87	31.52
19	8.91	10.12	11.65	30.14	32.85
20	9.59	10.85	12.44	31.41	34.17
21	10.28	11.59	13.24	32.67	35.48
22	10.98	12.34	14.04	33.92	36.78
23	11.69	13.09	14.55	35.17	38.07
24	12.40	13.85	15.66	36.41	39.36
25	13.12	14.61	16.47	37.65	40.65
26	13.84	15.38	17.29	38.88	41.92

where χ^2 is the chi-square, $\alpha = 1$ – confidence level $= 1 - 0.95 = 0.05$, and

$$\chi^2\left[\left(1-\frac{\alpha}{2}\right),(m-1)\right] = \chi^2\left[\left(1-\frac{0.05}{2}\right),(25-1)\right]$$

$$= \chi^2[0.975, 24]$$

$$= 12.4$$

As the value of S_{b25} (i.e., 18.28) lies between the above two values (i.e., 39.36 and 12.4), there is no contradiction to the assumption of exponential distribution.

7.3.2 Test II

This is another quite useful test that can also be used to determine whether the failure data follow an exponential distribution. The test call for computing the value of A, the χ^2 variate with $2m$ degrees of freedom. A is expressed by [6, 7, 12]

$$A = -2\sum_{i=1}^{m} \ln\left[\frac{T(\tau_i)}{T(\tau)}\right] \tag{7.4}$$

where

m = total number of failure times in a given sample
$T(\tau_i)$ = total operating time at the occurrence of failure i
$T(\tau)$ = total operating time at the termination of the test

If the value of A falls within

$$\chi^2\left[\frac{\alpha}{2}, 2m\right] < A < \chi^2\left[\left(1-\frac{\alpha}{2}\right), 2m\right] \tag{7.5}$$

then the assumption of an exponential distribution is not contradicted. The α is the risk of rejecting a true assumption and is expressed by

$$\alpha = 1 - \text{confidence level} \tag{7.6}$$

The application of this method is demonstrated through the following example:

Example 7.2

A total of 30 identical engineering items were tested for 300 h, out of which six items failed. None of the failed items were replaced. The failure times of the failed items are given in Table 7.3. Determine by using the above test at 95% confidence level that Table 7.3 data belong to an exponential distribution.

Total operating time of items at the termination of test

$$= (30 - 6)\,300 + (10 + 20 + 25 + 40 + 50 + 65)$$

$$= 7{,}410 \text{ h}$$

TABLE 7.3

Engineering Items' Times to Failure

No.	Item Failure Time (h)
1	10
2	20
3	25
4	40
5	50
6	65

Using the above result and the given data in Equation (7.3) yields

$$A=-2\left[\ln\left\{\frac{(10)(30)}{7410}\right\}+\ln\left\{\frac{10+(20)(29)}{7410}\right\}+\ln\left\{\frac{30+(25)(28)}{7410}\right\}\right.$$
$$\left.+\ln\left\{\frac{55+(40)(27)}{7410}\right\}+\ln\left\{\frac{95+(50)(26)}{7410}\right\}+\ln\left\{\frac{145+(65)(25)}{7410}\right\}\right]$$
$$=26.065$$

For a 95% confidence level using Equations (7.5–7.6), and the given data, we get

$$\alpha = 1 - 0.95 = 0.05$$

and

$$\chi^2[0.025,2(6)],\chi^2[0.975,2(6)]$$

Thus from Table 7.2, we get

$$\chi^2[0.025,12]=23.34 \quad \text{and} \quad \chi^2[0.975,12]=4.4$$

The value of A does not lie between the above two critical limits (i.e., 23.34 and 4.4). It means that the assumption of exponential distribution is contradicted.

7.4 Confidence Limit Estimation on Mean Item Life

As the exponential distribution is probably the most widely used distribution to represent failure behaviour of engineering items, this section is concerned with establishing confidence limits on mean time between failures (MTBF) when times to failures are exponentially distributed. The upper and lower confidence limits indicate the value of item MTBF being in an interval with certain confidence probability. More specifically, we may simply say that after testing a given sample of identical items for t h m failures occur; the item MTBF with certain confidence lies between specific upper and lower limits. To compute confidence limits on MTBF, the χ^2 (chi-square) distribution is utilized.

Two methods for establishing confidence limits on the MTBF of exponential distribution are presented below [4, 13].

7.4.1 Method I

This method is used when items are tested until the occurrence of a number of pre-assigned failures. Thus, the formulas for establishing one-sided (i.e., lower limit) and two-sided (i.e., upper and lower limits) confidence limits on MTBF, respectively, are as follows [4]:

$$\left[\frac{2t}{\chi^2(\alpha,2m)},\infty\right] \tag{7.7}$$

and

$$\left[\frac{2t}{\chi^2\left(\frac{\alpha}{2},2m\right)},\frac{2t}{\chi^2\left(1-\frac{\alpha}{2},2m\right)}\right] \tag{7.8}$$

where α is the probability that the interval will not contain the true value of MTBF and is expressed by $\alpha = 1$ – confidence level, and m is the number of failures.

The value of t is estimated by using either of the following two equations:

$$t \equiv kT \text{ (for replacement tests, i.e., failed items are replaced or repaired)} \tag{7.9}$$

and

$$t = \sum_{i=1}^{m} T_i + (k-m)T \text{ (for nonreplacement tests, i.e., in cases when failed items are not replaced or repaired)} \tag{7.10}$$

where

k = total number of identical items placed on test at time zero

T = time at the termination of life test

T_i = failure time of item i; for $i = 1, 2, 3, \ldots, m$

Example 7.3

Assume that a sample of 30 identical engineering items were tested until the occurrence of the tenth failure. Each failed item was replaced or repaired and the last failure occurred at 200 h. Compute the value of the one-sided (lower) confidence limit (i.e., the minimum value of MTBF of items) at 95% confidence level.

Using the given data in Equation (7.9) yields

$$t = (30)(200)$$
$$= 6000 \text{ h}$$

Similarly, the acceptable risk of error is

$$\alpha = 1 - (\text{confidence level})$$
$$= 1 - 0.95$$
$$= 0.05$$

Thus, using the above calculated value and the given data in Equation (7.7) and then Table 7.2 yields

$$\left[\frac{2(6000)}{\chi^2\{0.05, 2(10)\}}, \infty\right] = \left(\frac{12000}{31.41}, \infty\right) = (382.04, \infty)$$

It means, at 95% confidence level the minimum value of items' MTBF (i.e., mean life) is 382.04 h.

7.4.2 Method II

This method is used when the testing of a sample of items is terminated at a pre-assigned number of test hours. Thus, the formulas for establishing one-sided (i.e., lower limit) and two-sided (i.e., upper and lower limits) confidence limits on MTBF, respectively, are as follows [4]:

$$\left[\frac{2t}{\chi^2(\alpha, 2m+2)}, \infty\right] \tag{7.11}$$

and

$$\left[\frac{2t}{\chi^2\left(\frac{\alpha}{2}, 2m+2\right)}, \frac{2t}{\chi^2\left(1-\frac{\alpha}{2}, 2m\right)}\right] \tag{7.12}$$

The symbols t, α, and m are defined in the previous section.

Example 7.4

Assume that a sample of 30 identical engineering items was put on a test at time $t = 0$. The test was terminated at 200 h and the failed items were never replaced. During the test period a total of seven items failed at 50, 60, 85,

100, 120, 130, and 150 h. Compute the value of the one-sided (lower) confidence limit (i.e., the minimum value of MTBF or mean life of items) at 95% confidence level.

Using the given data in Equation (7.10) yields

$$\begin{aligned} t &= 50+60+85+100+120+130+150+(30-7)(200) \\ &= 5{,}295 \text{ h} \end{aligned}$$

Similarly, the acceptable risk of error is

$$\begin{aligned} \alpha &= 1-(\text{confidence level}) \\ &= 1-0.95 \\ &= 0.05 \end{aligned}$$

Thus, using the above two calculated values and the specified data in Equation (7.11) and then Table 7.2, we get

$$\begin{aligned} \left[\frac{2(5{,}295)}{\chi^2(0.05, 2(7)+2)}, \infty\right] &= \left[\frac{10{,}590}{\chi^2(0.05, 16)}, \infty\right] \\ &= \left(\frac{10{,}590}{26.29}, \infty\right) \\ &= (402.81, \infty) \end{aligned}$$

Thus, the minimum value of the engineering items' MTBF is 402.81 h at 95 % confidence level.

Example 7.5

Using Example 7.4 data, compute the items' MTBF (mean life) and its upper and lower confidence limits at 95% confidence level.

Dividing the calculated value of t in Example 7.4 by the total number of failures, the items' estimated MTBF is

$$\begin{aligned} \hat{\theta} &= \frac{5{,}295}{7} \\ &= 756.43 \text{ h} \end{aligned}$$

where $\hat{\theta}$ is the items' estimated MTBF.

Using the given and other data in Equation (7.12) and then utilizing Table 7.2, we get the following lower limit value of MTBF:

$$\text{MTBF}_L = \frac{2(5295)}{\chi^2(0.025, 16)} = \frac{10590}{28.84} = 367.20 \text{ h}$$

where MTBF_L is the lower limit value of MTBF.

Similarly, we get

$$\text{MTBF}_U = \frac{2(5295)}{\chi^2(0.975,14)} = \frac{10590}{5.63} = 1880.99 \text{ h}$$

where MTBF_U is the upper limit value of MTBF.

Thus, the item MTBF will lie between 367.20 and 1880.99 h with 95% confidence and its estimated expected value is 756.43 h.

7.5 Economics of Testing

Just like in the case of any other engineering activity, cost plays an important role in testing. Particularly, when determining the number of incoming items to be inspected or tested with respect to cost. One simple approach for achieving this is to first equate the cost of testing with the cost of not-testing and then determine the number of items to be tested or inspected at this break-even point. Mathematically, this is demonstrated as follows [14]:

The cost of testing is expressed by

$$\text{CT} = \text{CTE} + (nR)/m \tag{7.13}$$

where

CT = cost of testing
CTE = cost of the test equipment
n = number of items under consideration for testing or to be tested
R = labour and overhead rate per test hour
m = total number of items tested per hour

Similarly, the cost of not-testing is expressed by

$$\text{CNT} = n\,(\text{TFD})[(\text{FDP})(\text{CIR}) + (\text{FDW})(\text{CWR})] \tag{7.14}$$

where

CNT = cost of not-testing
TFD = total fraction defective previously estimated or observed
FDP = fraction defective which fails in-house or in-plant
CIR = average in-plant repair cost
FDW = fraction defective which fails during the warranted period
CWR = average warranty repair cost

To break-even, we equate the left-hand sides of Equation (7.13) and Equation (7.14):

$$\text{CT} = \text{CNT} \tag{7.15}$$

Substituting the right-hand sides of Equation (7.13) and Equation (7.14) into Equation (7.15) yields

$$\mathrm{CTE} + \frac{nR}{m} = n(\mathrm{TFD})[(\mathrm{FDP})(\mathrm{CIR}) + (\mathrm{FDW})(\mathrm{CWR})] \tag{7.16}$$

By solving Equation (7.16) in the term of n, we get

$$n^* = \frac{\mathrm{CTE}}{\mathrm{TFD}[(\mathrm{FDP})(\mathrm{CIR}) + (\mathrm{FDW})(\mathrm{CWR})] - \dfrac{R}{m}} \tag{7.17}$$

where $n*$ is the number of items to be tested (inspected) for break-even cost.

7.6 Problems

1. Discuss three important types of reliability testing.
2. What is the Bartlett test?
3. Define the following two terms:
   - Confidence limits
   - Confidence level
4. Discuss the following two terms:
   - Mean life
   - Chi-square distribution
5. Write an essay on reliability testing.
6. An engineering system was closely monitored with respect to failures over a period of time. A sample of its 30 times to failure (in days) is given in Table 7.4. Determine by using the Bartlett test at 90% confidence level that Table 7.4 data follow an exponential distribution.
7. A sample of 25 identical components were tested until the occurrence of the seventh failure. Each failed component was replaced and the last failure occurred at 150 h. Compute the value of the

TABLE 7.4

An Engineering System's Times to Failure (in Days)

10	60	110	170	210	260
28	75	120	175	213	270
40	90	125	190	230	275
50	100	140	200	240	280
55	105	150	205	250	300

one-sided (lower) confidence limit (i.e., the minimum value of the component MTBF) at 95% confidence level.

8. Using Question 7 data, compute the components' MTBF (mean life) and its upper and lower limits at 90% confidence level.
9. Assume that a sample of 40 identical components was put on a test at time $t = 0$. The test was terminated at 250 h and the failed components were never replaced. During the test period a total of six components failed at 80, 95, 140, 150, 160, and 170 h. Compute the value of the one-sided (lower) confidence limit (i.e., the minimum value of the component MTBF) at 95% confidence level.
10. Using Question 9 data, compute the components' MTBF (mean life) and its upper and lower limits at 90% confidence level.

References

1. Dhillon, B.S., *Reliability and Quality Control: Bibliography on General and Specialized Areas*, Beta Publishers, Gloucester, ON, 1992.
2. MIL-HDBK-781, Reliability Test Methods, Plans and Environments for Engineering Development, Qualification, and Production, U.S. Department of Defense, Washington, D.C.
3. MIL-STD-781, Reliability Design Qualification and Production Acceptance Test: Exponential Distribution, U.S. Department of Defense, Washington, D.C.
4. Von Alven, W.H., Ed., *Reliability Engineering*, Prentice Hall, Englewood Cliffs, NJ, 1964.
5. Dhillon, B.S., *Design Reliability: Fundamentals and Applications*, CRC Press, Boca Raton, FL, 1999.
6. Epstein, B., Tests for the validity of the assumption that the underlying distribution of life is exponential, *Technometrics*, Vol. 2, 1960, pp. 167–183.
7. Epstein, B., Tests for the validity of the assumption that the underlying distribution of life is exponential, *Technometrics*, Vol. 2, 1960, pp. 83–101.
8. Kapur, K.C. and Lamberson, L.R., *Reliability in Engineering Design*, John Wiley & Sons, New York, 1977.
9. Lamberson, L.R., An evaluation and comparison of some tests for the validity of the assumption that the underlying distribution of life is exponential, *AIIE Transactions*, Vol. 12, 1974, pp. 327–335.
10. Mann, N.R., Schafer, R.E., and Singpurwalla, N.D., *Methods for Statistical Analysis of Reliability and Life Data*, John Wiley & Sons, New York, 1974.
11. AMCP 702-C, *Quality Assurance: Reliability Handbook, Headquarters*, U.S. Army Material Command, Washington, D.C., October 1968.
12. NAVWEPS-00-65-502, *Handbook: Reliability Engineering*, Bureau of Naval Weapons, Department of Navy, Washington, D.C., 1964.
13. Dhillon, B.S., *Quality Control, Reliability, and Engineering Design*, Marcel Dekker, New York, 1985.
14. Kohoutek, H.J., Economics of reliability, in *Handbook of Reliability Engineering and Management*, Grant-Ireson, W. and Coombs, C.F., Eds. McGraw-Hill, New York, 1988, pp. 3.1–3.24.

8

Reliablity Management and Costing

8.1 Introduction

Reliability management is concerned with the direction and control of the reliability-related activities of the organization including establishing reliability goals and policies, staffing, and facilitating interactions of reliability manpower with the other parts of the organization.

The history of reliability management may be traced back to late 1950s when the Ballistic Missile Division of the U.S. Air Force developed a reliability program document (i.e., Exhibit 58-10) [1]. Subsequently, the U.S. Department of Defense developed the military specification, MIL-R-27542. This was the result of Department's efforts to develop requirements for an organized contractor reliability program. In 1985, a book entitled "Reliability and Maintainability Management" was published [2]. Needless to say, over the years a large number of publications on reliability management have appeared and a comprehensive list of these publications is given in Ref. [3].

Cost is an important factor in any reliability program. It is associated with various reliability-related activities including predication, allocation, and testing [4]. Reliability cost directly or indirectly is an important component of the item/product life cycle cost and over the years many publications directly or indirectly relating to it have appeared [5, 6].

This chapter presents various important aspects of reliability management and costing.

8.2 General Management Reliability Program-Related Responsibilities and Guiding Force Associated Facts for an Effective Reliability Program

General management plays a key role in having an effective reliability program. Some of its important responsibilities are shown in Figure 8.1 [7]. The responsibility "Establish reliability goals" is concerned with developing most

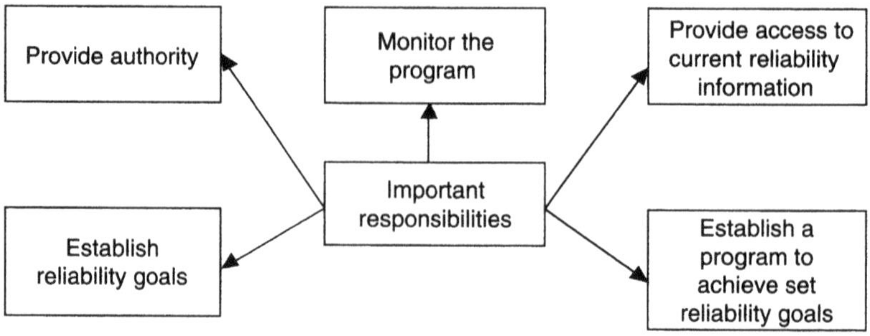

FIGURE 8.1
Reliability program-related responsibilities of the general management.

appropriate and achievable reliability goals for the organization. The responsibility "Provide access to current reliability information" is concerned with developing an effective mechanism, to access with respect to reliability information on current performance of company operations and the product/system/item it manufactures.

The responsibility "Provide authority" is concerned with providing appropriate reliability program-related authority, manpower, funds, and time schedule. The responsibility "Establish a program to achieve set reliability goals" is concerned with developing a mechanism to achieve set reliability goals and eradicating altogether the existing deficiencies. This mechanism or program, if effective, should be able to pay in return many times its establishing cost.

The responsibility "Monitor the program" is concerned with regularly monitoring the reliability program and then modifying the associated procedures, policies, organization, etc. to the most effective level.

Some of the facts that can be a guiding force for the general management to have an effective reliability program, are as follows [8]:

- Reliability is established by the basic design.
- Product reliability can only be improved through design changes.
- Reliability is a critical element in the planning, management, and design of an engineering item.
- For controlling reliability in design, manufacturing, testing, and field phases of the engineering product, properly planned programs are absolutely essential.
- Design reliability is lowered by the changes in manufacturing, storage and shipping, maintenance, testing and usage in the field of the engineering product.
- Product reliability can be improved most economically during the early phases of the design and evaluation testing programs.

- For achieving the desired reliability in mature engineering products in a timely manner, deficiency data collection, analysis, and feedback are essential.
- The occurrence of human errors degrades the reliability of design.

8.3 An Approach for Developing Reliability Goals and Useful Guidelines for Developing Reliability Programs

The development of reliability goals is absolutely essential when working from pre-established reliability requirements. It involves reducing the overall requirements to a series of sub goals. Nonetheless, an approach/procedure for developing reliability goals is presented below [9].

- Clarify requirements and review the organizational goal and the organizational unit's mission before establishing unit's goals to pursue the overall or main goal.
- Identify desired result areas and determine areas in which highest payoff looks promising.
- Select most promising result areas for pursuance and select appropriate goal candidates.
- Review resource needs for pursuing and achieving candidate goals successfully.
- Identify all possible problem areas in achieving the goals and proposed solutions.
- Rank candidate goals by considering the degree of payoff and the ease of payment.
- Review independence of goals and make appropriate changes in goal candidates to achieve the maximum level of coordination.
- Review the candidate goals with respect to factors such as measurability, acceptability, compatibility, supportability, relevance, and acceptability, and make final selection of goals and establish appropriate milestones for their successful achievement by developing success measurability parameters.
- Develop action plans for achieving the goals by considering factors such as management support and provisions for motivational initiatives/supervisions.
- Communicate goals in writing to all concerned people.
- Review goal progress on a regular basis and make adjustments accordingly.

Some of the useful guidelines for developing reliability programs are as follows [10]:

- Assign reliability-related goals for product/system under consideration.
- To obtain maximum "inherent product reliability," put in maximum effort during the design phase.
- Evaluate reliability margins and conduct specification review.
- Conduct procedure and design reviews.
- Develop a testing program.
- Review specification and drawing changes with respect to reliability.
- Establish and maintain control during production through activities such as sample testing, inspection, and effective production control.
- Establish a closed-loop system for failure reporting, analysis, and feedback to engineering professionals for appropriate actions to eliminate re-occurrence.
- Assign responsibility for reliability to the most appropriate group.
- Ensure that for the effective performance of reliability activities, the reliability group is reporting to an appropriate authority.
- Establish an appropriate on-the-job training facility and program.

8.4 Reliability Engineering Department Responsibilities and Tasks of a Reliability Engineer

A reliability engineering department performs various types of reliability-related functions. Some of its important responsibilities are shown in Figure 8.2 [11].

A reliability engineer performs various types of tasks during planning, design and development, manufacturing, and improvement phases of a system. These tasks include analyzing the proposed design with respect to reliability, developing reliability program plans, investigating field failures, securing appropriate resources for an effective reliability program during the planning stage, evaluating reliability of alternative designs, participating in design reviews, providing relevant reliability information to management, participating in evaluating request for proposals, developing reliability prediction models and methods, analyzing customer complaints with respect to reliability, budgeting/allocating the tolerable system failure down to the component level, running reliability-associated tests on the system, subsystems,

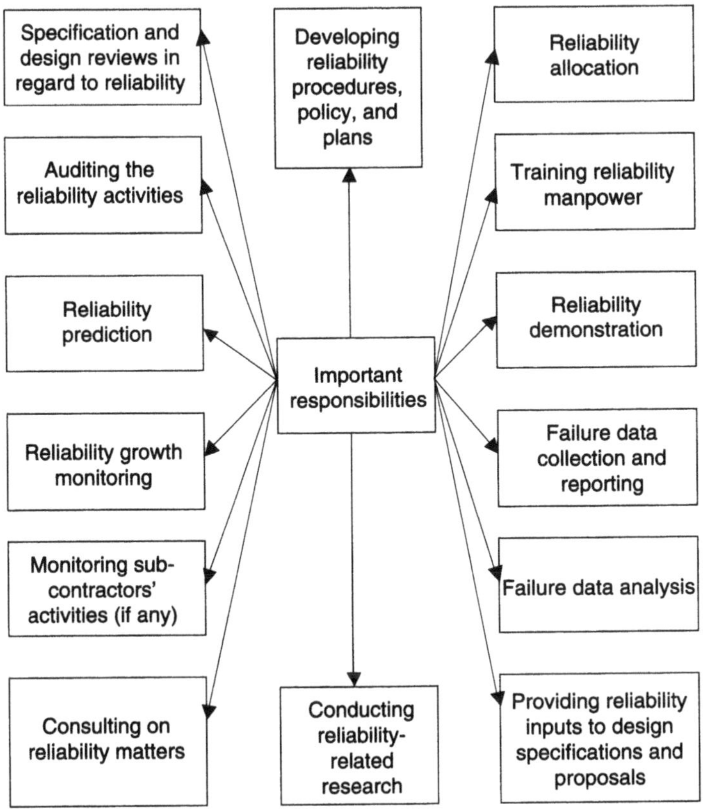

FIGURE 8.2
Important responsibilities of a reliability engineering department.

and parts, monitoring the reliability performance of subcontractors, keeping track of parts, reviewing difficulties associated with maintaining a system, participating in developing requirements for new systems, ensuring that the reliability relevant information flows to all concerned people, providing information to designers regarding improving the ease of maintenance and system life, developing tests on the system, subsystems, and components, and ensuring that appropriate consideration is given to reliability in new contracts [11].

8.5 Reliability Cost

This cost may be divided into four distinct categories as shown in Figure 8.3 [6].

The prevention cost includes items such as hourly cost and overhead rates for reliability engineers, design engineers, technicians, material engineers, and test and evaluation personnel; cost of preventive maintenance programs,

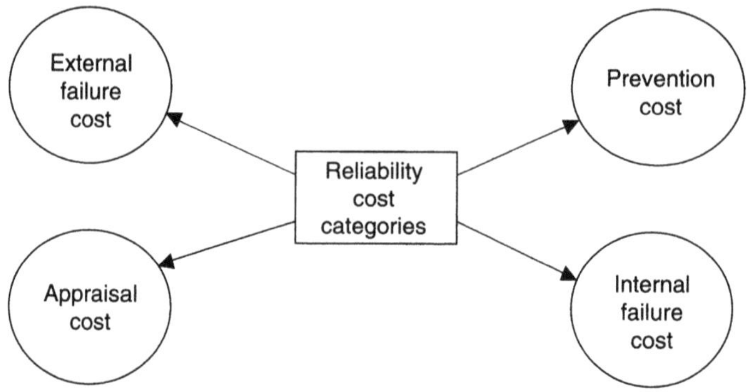

FIGURE 8.3
Reliability cost categories.

hourly cost and overhead rates for reliability screens, and cost of annual reliability training per capita.

The appraisal cost includes items such as hourly and overhead rates for reliability evaluation, reliability demonstration, life testing, reliability qualification, and environmental testing; vendor assurance cost for new part qualification, new vendor qualifications, and vendor audit; average cost per part of assembly testing, screening, auditing, calibration, and inspection; and cost of test result reports.

The external failure cost includes items such as failure analysis cost, cost to repair a failure, cost of replaced parts, cost of liability assurance, service engineering hourly rate and overhead, warranty administration and reporting cost, cost of service kits, and cost of spare part inventory.

The internal failure cost includes items such as cost of replaced parts, cost of spare part inventory, cost of production change administration, and hourly cost and overhead rates for troubleshooting and repair, failure analysis, and retesting.

8.6 Reliability Activity Cost Estimation Models

A reliability program involves the performance of various activities or tasks. Some examples of these tasks are reliability prediction, reliability modeling/allocation, preparing the reliability and maintainability program plan, reliability testing, and performing failure modes and effect analysis (FMEA). Manhours required to perform tasks such as these can be estimated by using various empirically-based mathematical models [4, 12] and in turn, the cost of performing such tasks.

8.6.1 Reliability Prediction Manhour Estimation Model

Reliability prediction is one of the most important tasks of a reliability program. The total number of manhours required to perform this task can be estimated by using the following equation [4]:

$$MH_{rp} = 4.54(POC)(LOD)^2\,(RF)^2 \tag{8.1}$$

where

MH_{rp} = total number of manhours required for reliability prediction.

POC = denotes the percentage of commercial hardware used in system/item under consideration. The recommended values of the POC are 1 (76–100%), 2 (51–75%), 3 (26–50%), and 4 (0–25%).

LOD = denotes the level of detail. The recommended values of the LOD are 1 (if prediction exists), 2 (if the prediction is to be made using similar system data), and 3 (if full MIL-HDBK-217 [13] stress prediction is required).

RF = denotes the report formality (i.e., the type of report required). The recommended values of the RF are 1 (if internal report is required), and 2 (if formal report is required).

8.6.2 Reliability Modeling/Allocation Manhour Estimation Model

This model can be used to estimate manhours required to perform reliability modeling/allocation. The total number of manhours required to perform this task is expressed by [4]:

$$MH_{rma} = 4.05\,(MAAC)^2\,(NOIIAP) \tag{8.2}$$

where

MH_{rma} = total number of manhours required for reliability modeling/allocation.

MAAC = denotes the modeling and allocation complexity. The recommended values of the MAAC are 1 (if a series system), 2 (if simple redundancy), and 3 (if very complex redundancy).

NOIIAP = denotes the number of items in allocation process. The recommended minimum value of NOIIAP is 7 and the maximum 445.

8.6.3 Reliability and Maintainability Program Plan Manhour Estimation Model

This model is concerned with estimating the number of manhours required to prepare the reliability and maintainability program plan. The total number of manhours required to perform this task is expressed by [4]

$$MH_{rmp} = 2.073\,\alpha^2 \tag{8.3}$$

where MH_{rmp} is the total number of manhours required for preparing the reliability and maintainability program plan and α is the total number of MIL-STD-785/470 [14] tasks required. The recommended minimum value of α is 4 and the maximum 22.

8.6.4 Reliability Testing Manhour Estimation Model

This model is concerned with estimating the total number of manhours required for performing the reliability testing task. The total number of manhours required to perform the task is expressed by [4]

$$MH_{rt} = (182.07)(HC) \tag{8.4}$$

where MH_{rt} is the total number of manhours required for reliability testing and HC denotes the hardware complexity. The recommended values of the HC are 1 (if the parts are less than 15,000), 2 (if the parts are 15,000–25,000), and 3 (if the parts are greater than 25,000).

8.6.5 Failure Modes and Effect Analysis (FMEA) Manhour Estimation Model

This model can be used to estimate manhours required to perform FMEA. The total number of manhours required to perform this task is expressed by [4]:

$$MH_{FM} = 17.79(NUI) \tag{8.5}$$

where MH_{FM} is the total number of manhours required for performing FMEA. NUI is the number of unique items requiring FMEA (e.g., quantity of equipment for equipment level FMEA or number of circuit cards for piece part and circuit level FMEA). The recommended minimum value of NUI is 3 and the maximum 206.

8.6.6 Failure Reporting and Corrective Action System (FRACAS) Manhour Estimation Model

This model is concerned with estimating the total number of manhours required for FRACAS. The total number of manhours required is expressed by [4]

$$MH_{FR} = 8.25(DOFI)^2 \tag{8.6}$$

where MH_{FR} is the total number of manhours required for FRACAS and DOFI is the duration of FRACAS implementation expressed in months. The recommended minimum value of the DOFI is 2.5 and the maximum 38.

8.6.7 Reliability and Maintainability Management Manhour Estimation Model

This model is concerned with estimating the manhours required to perform the activity of reliability and maintainability management. The total number of manhours required is expressed by [4]:

$$MH_{RM} = [MH_{rp} + MH_{rma} + MH_{rmp} + MH_{rt} + MH_{FM} + MH_{FR}]\left[\frac{16}{100}\right] \quad (8.7)$$

where MH_{RM} is the total number of manhours required for performing the task of reliability and maintainability management.

8.7 Problems

1. Write an essay on the history of reliability management.
2. Discuss important reliability-related responsibilities of general management.
3. Discuss the facts that can be a guiding force for the general management to have an effective reliability program.
4. Discuss a useful approach for developing reliability goals.
5. List at least ten useful guidelines for developing a reliability program.
6. What are the important responsibilities of a reliability engineering department?
7. Discuss the following two categories of the reliability cost:
   - Appraisal cost
   - Prevention cost
8. What are the advantages and disadvantages of the empirically-based, reliability program activity, cost estimation models?
9. Discuss two important reliability goals.
10. List four categories of the reliability cost.
11. What are the principal tasks of a reliability engineer?

References

1. Austin-Davis, W., Reliability management: a challenge, *IEEE Trans. on Reliability*, Vol. 12, 1963, pp. 6–9.
2. Dhillon, B.S. and Reiche, H., *Reliability and Maintainability Management*, Van Nostrand Reinhold, New York, 1985.

3. Dhillon, B.S., *Reliability and Quality Control: Bibliography on General and Specialized Areas*, Beta Publishers, Gloucester, ON, 1992.
4. RADC Reliability Engineer's Toolkit, Published by the Systems Reliability and Engineering Division, Rome Air Development Center (RADC), Air Force Systems Command (AFSC), Griffiss Air Force Base, Rome, New York, July 1988.
5. Dhillon, B.S., *Life Cycle Costing: Techniques, Models, and Applications*, Gordon and Breach Science Publishers, New York, 1989.
6. Grant Ireson, W., Coombs, C.F., Ed., *Handbook of Reliability Engineering and Management*, McGraw-Hill, New York, 1988.
7. Heyel, C., *The Encyclopaedia of Management*, Van Nostrand Reinhold, New York, 1979.
8. Finch, W.L., Reliability: a technical management challenge, *Proceedings of the American Society for Quality Control Annual Conference*, 1981, pp. 851–856.
9. Grant Ireson, W., Coombs, C.F. and Moss, R.Y., Eds., *Handbook of Reliability Engineering and Management*, McGraw-Hill, New York, 1996.
10. Karger, D.W. and Murdick, R.G., *Managing Engineering and Research*, Industrial Press, New York, 1980.
11. Dhillon, B.S., *Design Reliability: Fundamentals and Applications*, CRC Press, Boca Raton, FL, 1999.
12. R and M Program Cost Drivers, Report No. RADC-TR-87-50 (ADA 182773). Available from the National Technical Information Service (NTIS), Springfield, VA.
13. MIL-HDBK-217, *Reliability Prediction of Electronic Equipment*, Department of Defense, Washington, D.C.
14. MIL-STD-785, *Reliability Program for Systems and Equipment*, Department of Defense, Washington, D.C.

9

Introduction to Quality

9.1 Introduction

The importance of quality in business and industry is increasing rapidly. Today, our daily lives and schedules are more dependent than ever before on the satisfactory functioning of products and services (e.g., computers, automobiles, and a continuous supply of electricity). Needless to say, factors such as competition, product sophistication, growing demand from customers for better quality, increasing number of quality-related lawsuits, and the global economy have played an instrumental role in increasing the importance of quality.

The cost of quality control accounts for approximately 7–10% of the total sales revenue of manufacturers [1]. Some of the quality-related challenges facing the industry are a very high level of cost for quality, an alarming rate of increase in customer quality requirements, need for improvements in methods and practices associated with quality-related activities, and the Internet economy.

This chapter presents important introductory aspects of quality.

9.2 Comparisons of Modern and Traditional Products, Direct Factors Influencing the Quality of Product and Services, and Quality Design Characteristics

Modern products differ quite considerably from the traditional ones. Table 9.1 presents a comparison of modern and traditional products with respect to various factors [2].

The important factors that directly influence the quality of products and services are as follows [1, 3]:

- Machines used in manufacturing
- Modern information methods

TABLE 9.1

Comparisons of Modern and Traditional Products with Respect to Various Factors

No.	Factor	Modern Product	Traditional Product
1	Understanding of product by users	Low	High
2	Production volume	Usually high	Usually low
3	Precision	High	Low
4	Simplicity	Complex and dynamic	Simple and dynamic
5	Design life	Short (usually less than ten years)	Long (decades)
6	Use environment	Unnatural	Natural
7	Importance to human health and safety	Often important	Seldom important
8	Requirement for interchange ability	Usually extensive	Usually limited
9	Scientific basis of design	Generally scientific	Basically empirical
10	Frequent causes of field failures	Design shortcomings	Manufacturing errors
11	Life cycle cost to users	Much higher than procurement cost	Similar to procurement cost
12	Basis of reliability and maintainability	Quantified	Vague

- Money, manpower, and materials
- Mounting product requirements
- Motivation of employees
- Market for product and services
- Management

Quality design characteristics, under the control of the producer, for both a manufactured and a service product are performance, features, conformance, reliability, durability, serviceability, and aesthetics [4]. Their examples for a manufactured product (i.e., Stereo amplifier) are signal-to-noise ratio, remote control, workmanship, mean time to failure, useful life, ease of repair, and oak cabinet, respectively. Similarly, their examples for a service product (i.e., checking account) are time to process customer requests, automatic bill paying, accuracy, variability of time to process requests, keeping pace with industry trends, resolution of errors, and appearance of bank lobby, respectively.

9.3 Quality Goals

Usually in organizations attainable goals for quality are developed first, and then efforts are directed to meet these objectives. Some organizations may group their quality goals under the following two categories [5]:

- Goals for breakthrough
- Goals for control

The goals for breakthrough are basically concerned with improving the existing quality of product or services. There could be various reasons for establishing such goals including enhancing company image in the market, dissatisfied customers and others with the present products or services, retaining or attaining quality leadership, and loosing market share because of failure to compete with similar products or services provided by others.

The goals for control are concerned with maintaining the quality of products or services to the existing level for a given period. Some of the reasons for such goals are: improvements are uneconomical, insignificant number of customers or other complaints about the quality of products or services, and acceptable competitiveness at present quality levels.

All in all, quality goals should be developed by following steps such as those listed below [4].

- Identifying potential goals
- Quantifying potential goals
- Setting goal priorities

9.4 Quality Assurance System Elements

The main objective of a quality assurance system is to maintain the required level of quality. Its important elements/tasks are shown in Figure 9.1 [6].

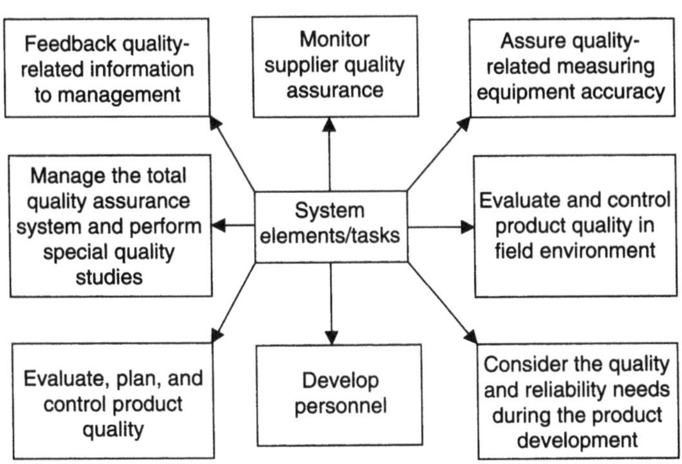

FIGURE 9.1
Quality assurance system elements/tasks.

9.5 Design for Quality

There are many aspects of product design that can considerably degrade quality. The design of a product affects quality in the following two major areas [3]:

- In the manufacturer's own plant.
- At the supplier's plant.

Figure 9.2 presents a number of design guidelines for quality improvement [7].

The guideline "minimize number of components" leads to results (along with their specific corresponding quality improvements) such as less complicated assemblies (lower assembly error rate), fewer part and assembly drawings (less volume of drawings and instructions), fewer components to fail (higher reliability), and fewer components to hold to required quality characteristics (highest consistency of component quality). The guideline "eliminate adjustments altogether" leads to results (along with their specific corresponding quality improvements) such as no assembly adjustment errors (higher first-pass yield) and eradication of adjustable parts with high failure rates (lower failure rate).

The guideline "design for robustness" leads to (along with its specific corresponding quality improvements) low sensitivity to component variability (less degradation of performance over time and greater first-pass yield).

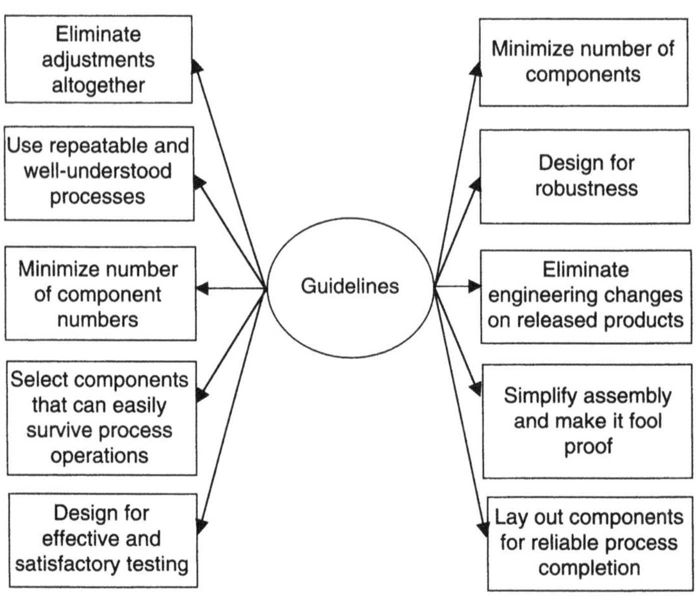

FIGURE 9.2
Design guidelines for improving product quality.

The guideline "minimize number of component numbers" lead to (along with its specific corresponding quality improvement) fewer variations of like components (lower assembly error rate).

The guideline "simplify assembly and make it foolproof" leads to results (along with their specific corresponding quality improvements) such as parts cannot be assembled wrong (lower assembly error rate), obvious visibility of missing parts (lower assembly error rate), self-securing parts (lower assembly error rate), and no "force fitting" of parts (less damage to parts and better service-ability). The guideline "use repeatable and well-understood processes" leads to results (along with their specific corresponding quality improvements) such as assembly quality easy to control (higher assembly yield) and part quality easy to control (higher part yield).

The guideline "design for effective and satisfactory testing" leads to (along with its specific corresponding quality improvements) less mistakes in identifying "good" and "bad" product and vice-versa (truer assessment of quality and less unnecessary rework). The guideline "select components that can easily survive process operations" leads to results (along with their specific corresponding quality improvements) such as less degradation of parts (higher reliability) and less damage to parts (higher yield). The guideline "eliminate engineering changes on released products" leads to (along with its specific corresponding quality improvement) fewer errors due to changeovers and multiple revisions/versions (lower assembly error rate). The guideline "lay out components for reliable process completion" leads to (along with its specific corresponding quality improvements) less damage to parts during handling and assembly (higher yield and higher reliability).

9.6 Total Quality Management (TQM)

The term total quality management (TQM) was coined by Nancy Warren, a behavioral scientist, in 1985 [8]. It is made up of three words, each of which is described in detail below.

1. **Total.** This calls for an effective team effort of all involved parties to satisfy customers. There are many factors that play an important role in developing a successful supplier-customer relationship. Some of the important ones are as follows:
   - Customers making suppliers understand their obligations or needs effectively.
   - Development of customer-supplier relationships on the basis of mutual trust and respect.
   - Monitoring of suppliers' products and processes by customers on a regular basis.
   - Customers developing their internal needs.

2. **Quality.** There are numerous definitions of quality. Nonetheless, quality must be viewed from the customer perspective. This factor is further reinforced by the result of a survey conducted by the Conference Board of Canada, in which over 80% of the respondents stated that quality is defined by the customer and not by the supplier [9].
3. **Management.** An effective approach to management is crucial in determining company ability to attain corporate objectives and to allocate resources in an effective manner. The TQM approach requires an effective involvement of employees in company decision making because their participation and contribution are viewed as critical to mold all areas of business in providing high quality products and services to customers.

In order to practice the TQM concept in an effective manner, it is essential to understand fundamental differences between TQM and the traditional quality assurance management (TQAM). Table 9.2 presents comparisons in a number of areas between TQM and TQAM [10, 11].

9.6.1 TQM Elements and Goals for TQM Process Success

TQM is composed of many elements. The important ones are shown in Figure 9.3 [12].

TABLE 9.2

Comparisons Between Total Quality Management and the Traditional Quality Assurance Management

No.	Area	Total Quality Management	Traditional Quality Assurance Management
1	Customer	A well-defined approach to comprehend and satisfy customer requirements	Ambiguous understanding of customer or consumer requirements
2	Decision making	Practiced an effective team approach with team of employees	Practiced usual top-down method
3	Quality defined	Products suitable for consumer applications	Products satisfy specifications
4	Objective	Prevent the occurrence of errors	Discover errors
5	Quality responsibility	All people in the organization involved	Quality control group/ inspection center
6	Cost	Better quality decreases cost and increases productivity	Improvements in quality result in higher cost
7	Definition	Customer-driven	Product-driven

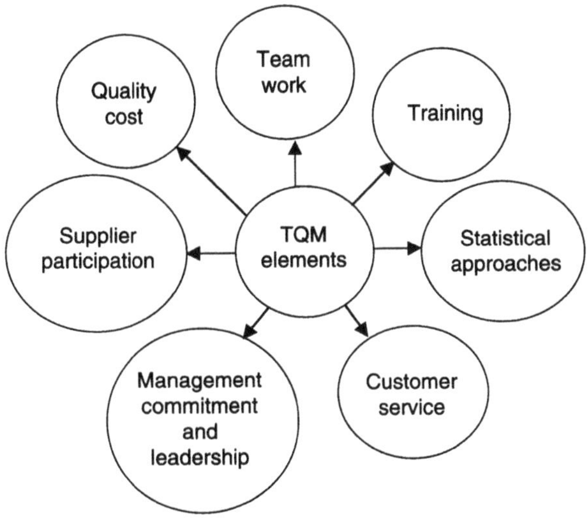

FIGURE 9.3
Important elements of TQM.

For TQM process success, there are a number of goals that must be fulfilled effectively. Some of these goals are as follows [13]:

- Clear understanding of internal and external customer requirements by all company employees.
- Establishment of rewards and incentives for employees when process control and customer satisfaction results are attained.
- Use of a system to continuously improve processes that better meet customers' present and future needs.
- Meeting of control guidelines per customer requirements by all concerned systems and processes.

9.6.2 Deming Approach to TQM

There have been many people who over the years directly or indirectly have contributed to TQM. One of these contributors was W.E. Deming, a graduate in engineering, mathematics, and physics. His fourteen step approach to improve quality is as follows [10, 14–16]:

- Develop constancy of purpose for enhancing services/products. More specifically, this requires the development of a mission statement addressing issues such as long term corporate objectives, quality philosophy, growth plans, investors, and employees.
- Lead to promote change. More specifically, this means that the existing acceptable levels of defects, delays, or mistakes are unacceptable and all concerned individuals/bodies are alerted to determine factors

for their (i.e., defects, delays, or mistakes) existence. Subsequently, everyone concerned work together to rectify highlighted problems.

- Stop depending on mass inspection and build quality into products/services.
- Stop awarding business or contracts on the basis of price and develop long-term relationships on the basis of performance
- Improve product, quality, and service continuously.
- Institute training measures that include modern/latest approaches, methods, and techniques
- Practice modern/latest supervisory approaches and methods.
- Eliminate the element of fear altogether.
- Break down existing barriers between groups/units/departments and emphasize team effort.
- Eliminate slogans, numerical goals, and posters because they create adversarial relationships.
- Eradicate numerical quotas and the practice of management by objectives (MBO).
- Eradicate all existing obstacles to employee pride in workmanship.
- Encourage dynamic education and self-improvement programs.
- Make the transformation everyone's task and force all concerned to work on it in an effective manner.

9.6.3 Obstacles to TQM Implementation

Over the years, individuals involved with TQM implementation have experienced many obstacles. Knowledge of these obstacles is considered important prior to embarking on the TQM implementation process. Table 9.3 presents some of these obstacles in the form of questions [17].

TABLE 9.3

Some TQM Obstacles in the Form of Questions

No.	Obstacle Related Question
1	Will upper management support the introduction of the TQM program?
2	Does management clearly understand TQM purpose?
3	Is it possible to quantify customer needs? If so, how?
4	Is it possible to obtain effective support of managers and their subordinates possessing an "independent" attitude?
5	Who will set the TQM vision?
6	How to convince individuals of the need to change?
7	Is there adequate time available for implementing TQM program in an effective manner?
8	How to convince all involved people that TQM is different?

9.6.4 Selected Books on TQM and Organizations that Promote the TQM Concept

Over the years many books on TQM and organizations that promote the TQM concept have appeared. This section lists some of these books and organizations separately.

Books

- Tenner, R.R. and Detoro, I.J., *Total Quality Management: Three Steps to Continuous Improvement*, Addison-Wesley, Reading, MA, 1992.
- Mizuno, S., *Company-Wide Total Quality Control*, Asian Productivity Organization, Tokyo, 1989.
- Shores, A.R., *Survival of the Fittest: Total Quality Control and Management*, ASQC Quality Press, Milwaukee, WI, 1988.
- Stein, R.E., *The Next Phase of Total Quality Management*, Marcel Dekker, New York, 1994.
- Gevirtz, C. D., *Developing New Products with TQM*, McGraw-Hill, New York, 1994.
- Walton, M., *The Deming Management Method*, Association for Quality and Participation, Cincinnati, Ohio, 1986.
- Spenley, P., *World Class Performance Through Total Quality*, Chapman Hall, London, 1992.
- Oakland, J.S., *Total Quality Management: Text with Cases*, Butterworth-Heinemann, Burlington, MA, 2003.
- Besterfield, D.H. et al. *Total Quality Management*, Prentice Hall, Upper Saddle River, NJ, 2003.
- Rampersad, H.K., *Total Quality Management*, Springer-Verlag, New York, 2000.

Organizations

- Quality and Productivity Management Association, 300 Martingale Road, Suite 230, Schamburg, IL.
- American Society for Quality Control, 611 East Wisconsin Avenue, P.O. Box 3005, Milwaukee, WI.
- American Productivity and Quality Center, 123 North Post Oak Lane, Houston, Tx.
- American Society for Training and Development, 1640 King Street, Alexandria, Va.

9.7 Problems

1. Discuss the importance of quality.
2. Discuss the factors that influence the quality of products and services.
3. Compare modern and traditional products.
4. Discuss quality goals.
5. List at least eight important elements/tasks of the quality assurance system.
6. List at least ten design guidelines for improving product quality.
7. Discuss the concept of total quality management (TQM).
8. Make a comparison between total quality management and traditional quality assurance management.
9. What are the important elements of TQM?
10. Describe the Deming approach to TQM.

References

1. Feigenbaum, A.V., *Total Quality Control*, McGraw-Hill, New York, 1983.
2. Juran, J.M. and Gryna, F.M., *Quality Planning and Analysis*, McGraw-Hill, New York, 1980.
3. Dhillon, B.S., *Quality Control, Reliability, and Engineering Design*, Marcel Dekker, New York, 1985.
4. Evans, J.R. and Lindsay, W.M., *The Management and Control of Quality*, West Publishing Company, New York, 1996.
5. Juran, J.M., Gryna, F.M., and Bingham, R.S., *Quality Control Handbook*, McGraw-Hill, New York, 1979.
6. The quality world of Allis-Chalmers, *Quality Assurance*, Vol. 9, 1970, pp. 13–17.
7. Daetz, D., The effect of product design on product quality and product cost, *Quality Progress*, June 1987, pp. 63–67.
8. Walton, M., *Deming Management at Work*, Putnam, New York, 1990.
9. Farquhar, C.R. and Johnston, C.G., Total Quality Management: A Competitive Imperative, Report No. 60-90-E, 1990. Available from the Conference Board of Canada, 255 Smyth Road, Ottawa, Ontario, Canada.
10. Schmidt, W.H. and Finnigan, J.P., *The Race without a Finish Line: America's Quest for Total Quality*, Jossey-Bass Publishers, San Francisco, 1992.
11. Madu, C.N. and Chu-hua, K., Strategic total quality management (STQM), in Madu, C.N., Ed., *Management of New Technologies for Global Competitiveness*, Quorum Books, Westport, CT, 1993, pp. 3–25.
12. Burati, J.L. Matthews, M.F., and Kalidindi, S.N., Quality management organization and techniques, *Journal of Construction Engineering and Management*, Vol. 118, March 1992, pp. 112–128.

13. Dhillon, B.S., *Advanced Design Concepts for Engineers*, Technomic Publishing Company, Lancaster, PA, 1998.
14. Goetsch, D.L. and Davis, S., *Implementing Total Quality*, Prentice Hall, Englewood Cliffs, NJ, 1995.
15. Mears, P., *Quality Improvement Tools and Techniques*, McGraw-Hill, New York, 1995.
16. Heizer, J. and Render, B., *Production and Operations Management*, Prentice Hall, Upper Saddle River, NJ, 1995.
17. Klein, R.A., Achieve total quality management, *Chemical Engineering Progress*, November 1991, pp. 83–86.

10

Quality Analysis Methods

10.1 Introduction

Just like in the case of other engineering disciplines, over the years many analytical methods for use in quality work have been developed. These methods are statistically and nonstatistically based [1, 2]. The main objective of all these methods is to improve the quality of products and services. The history of the quality analysis methods may be traced back to a memorandum written by Walter Shewhart of Bell Laboratories on May 16, 1924 [1, 3]. In this memorandum Walter Shewhart presented the idea of a control chart.

Today, there are many methods and techniques available in the published literature that can be used to perform various types of quality-related analysis. The effectiveness of these methods and techniques may vary from one application to another. It means a careful consideration must be given in their selection for use in specific applications with respect to factors such as cost, ease of use, the degree of accuracy required, and the particular phase in product design and manufacture. This chapter presents numerous methods that can be used to perform various types of quality-related analysis.

10.2 Quality Control Charts

Quality control charts are widely used in industry for various purposes including to obtain information whether the process is in the state of control or not and to provide information for decisions concerning inspection procedures or product specifications. A control chart may simply be described as a graphical approach used for determining whether a process is in a "state of statistical control" or out of control [4]. The construction of control charts is based on statistical principles and distributions. Figure 10.1 shows the basic form of a control chart. Basically, the chart is made up of three items: mean or standard value of the characteristic of interest, upper control limit (UCL), and lower control limit (LCL).

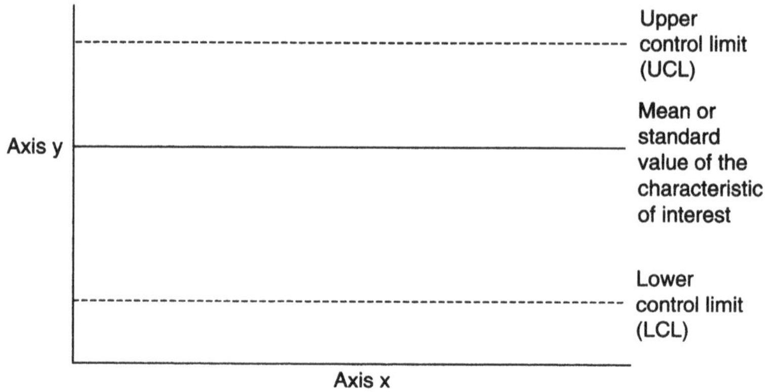

FIGURE 10.1
A rough sketch of a control chart.

Typically, random samples of values taken from a process are plotted on the control chart. When a sample value falls outside the upper and lower control limits, it signifies that the process is out of statistical control.

Although control charts were originally developed and used in manufacturing areas, they can easily be applied in service organizations. For example, in organizations (along with respective quality measures) such as hospital (laboratory test accuracy, on-time delivery of meals and medication, and insurance claim accuracy), post office (sorting accuracy, time of delivery, and percent express mail delivered on time), hotel (checkout time, number of complaints received, and proportion of rooms satisfactorily cleaned), bank (check-processing accuracy), police department (number of traffic citations and incidence of crime in a precinct), ambulance (response time), auto service (percent of time work completed as promised and number of complaints), and insurance company (billing accuracy and claims-processing response time) [4, 5].

There are many different types of control charts [6]. The most widely used charts are described below [1, 7, 8].

10.2.1 The *p*-Charts

Often, these charts are called the control charts for attributes. Attributes generally cannot be measured, but they can easily be counted. The data population is classified into two groups (i.e., good or bad, pass or fail, etc.).

An example of the data population is the components with defects and components without defects. The main advantage of the attributes data is usually easy to collect and the main disadvantage, the need to collect large samples to obtain valid statistical results. Finally, it may be said that attributes control charts use pass-fail information for charting and a *p*-chart simply is a single chart that tracks the proportion of nonconforming items in each and every sample.

The binomial distribution is used to establish upper and lower control limits of the *p*-chart. Thus, the upper and lower control limits are expressed by

$$\text{UCL}_{pc} = \mu_b + 3\sigma_b \tag{10.1}$$

and

$$\text{LCL}_{pc} = \mu_b - 3\sigma_b \tag{10.2}$$

where
UCL_{pc} = upper control limit of the *p*-chart
LCL_{pc} = lower control limit of the *p*-chart
μ_b = mean of the binomial distribution
σ_b = standard deviation of the binomial distribution

The mean μ_b is expressed by

$$\mu_b = \frac{n}{m\alpha} \tag{10.3}$$

where
α = total number of samples
m = sample size
n = total number of defectives (or failures, etc.) in classification

The standard deviation σ_b is expressed by

$$\sigma_b = [\mu_b(1 - \mu_b)/m]^{1/2} \tag{10.4}$$

Example 10.1
Assume that six samples were taken from the production line of a company manufacturing certain electrical parts. Each sample contained 50 parts. After a careful inspection, it was concluded that samples 1, 2, 3, 4, 5, and 6 contain 2, 4, 10, 6, 5, and 8 defective parts, respectively. Construct the *p*-chart for electrical parts.

By substituting the specified values into Equation (10.3), we get

$$\mu_b = \frac{(2+4+10+6+5+8)}{(50)(6)} = 0.117$$

Using the above calculated value and the other given data value in Equation (10.4) yields

$$\sigma_b = [0.117(1 - 0.117)/50]^{1/2}$$
$$= 0.045$$

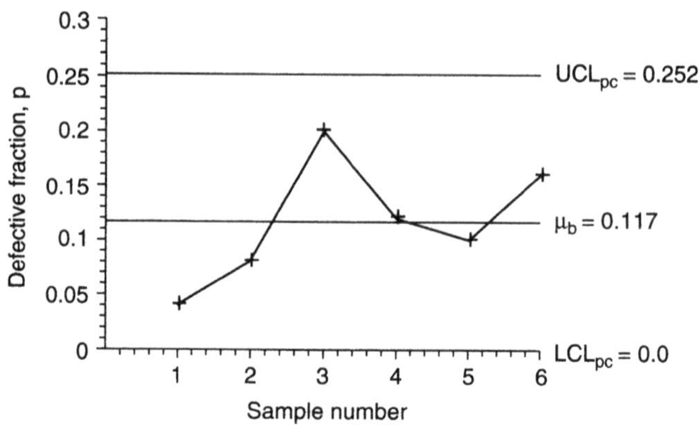

FIGURE 10.2
p-chart for electrical parts.

The fraction of defectives p in sample 1 is given by

$$p = \frac{2}{50} = 0.04$$

Similarly, the fraction of defective parts in samples 2, 3, 4, 5, and 6 are 0.08, 0.2, 0.12, 0.1, and 0.16, respectively.

Using the above calculated values for μ_b and σ_b in Equation (10.1) and Equation (10.2), we get

$$\text{UCL}_{pc} = 0.117 + 3(0.045) = 0.252$$

and

$$\text{LCL}_{pc} = 0.117 - 3(0.045) = -0.018 \cong 0$$

Figure 10.2 shows a *p*-chart for the above calculated values. The crosses in the figure denote the fraction of defective parts in each sample. As all the sample fractions are within the upper and lower control limits, it means that there is no abnormality in the production process.

10.2.2 The *R*-Charts

These charts are known as the control charts for ranges because they depict the variation in the ranges of the samples. The main reason for using ranges instead of the standard deviation is that they are much easier to compute by individuals, particularly on the shop floor. Nonetheless, the range of a sample is given by the difference between the highest and the lowest observation value in the sample. The sample ranges are plotted on the *R*-chart.

The upper and lower control limits for the R-chart are expressed by [5]

$$\mathrm{UCL}_R = \left[1 + \frac{3d_3}{d_2}\right]\overline{R} \tag{10.5}$$

and

$$\mathrm{LCL}_R = \left[1 - \frac{3d_3}{d_2}\right]\overline{R} \tag{10.6}$$

where

$\overline{R}$ = mean value of the sample ranges
UCL_R = upper control limit of the R-chart
LCL_R = lower control limit of the R-chart
d_2 and d_3 = factors whose values depend on the sample size n. These are tabulated in Table 10.1 (The values of these factors are calculated by assuming that the distribution of sample means is normal [5]).

The value of $\left[1 - \frac{3d_3}{d_2}\right]$ is taken as zero when a sample contains less than seven observations.

TABLE 10.1

Values for Factors d_2 and d_3

No.	n (sample size)	d_2	d_3
1	2	1.13	0.85
2	3	1.69	0.89
3	4	2.06	0.88
4	5	2.33	0.86
5	6	2.53	0.85
6	7	2.70	0.83
7	8	2.85	0.82
8	9	2.97	0.81
9	10	3.08	0.80
10	11	3.17	0.79
11	12	3.26	0.78
12	13	3.34	0.77
13	14	3.41	0.76
14	15	3.47	.076
15	16	3.53	0.75
16	17	3.59	0.74
17	18	3.64	0.74
18	19	3.69	0.73
19	20	3.74	0.73
20	21	3.78	0.72
21	22	3.82	0.72
22	23	3.86	0.72
23	24	3.90	0.71

TABLE 10.2

Observation Times for Eight Samples Labeled A to H

Observation No.	Sample A	B	C	D	E	F	G	H
1	25	15	30	15	23	15	18	14
2	10	13	14	26	15	7	25	28
3	15	9	25	15	24	10	22	9
4	8	15	15	18	26	15	23	16
5	10	18	12	20	14	8	20	14
6	12	20	10	13	25	14	15	19
7	16	23	18	18	13	11	21	17
8	18	10	16	16	10	9	14	15
9	12	11	21	20	11	13	13	13
10	15	19	11	10	19	16	13	16

TABLE 10.3

Samples A–H Ranges

	Samples A	B	C	D	E	F	G	H
Sample range	17	14	20	16	16	9	12	19

Example 10.2

A task is performed repetitively at a mechanical parts assembly line. Time taken to perform the task were collected randomly eight times over a period of time.

Each time, the task was observed ten times. Table 10.2 presents observed times at eight different occasions. Develop the *R*-chart.

Range for Sample A in Table 10.2 is

$$= \text{(Highest value in the sample)} - \text{(Lowest value in the sample)}$$

$$= 25 - 8 = 17$$

Similarly, ranges for other samples can be obtained. Table 10.3 presents ranges for samples A–H.

The mean $\bar{R}$ of the Table 10.3 ranges is

$$\bar{R} = 15.4$$

For a sample size n equal to ten, from Table 10.1 we get

$$d_2 = 3.08$$

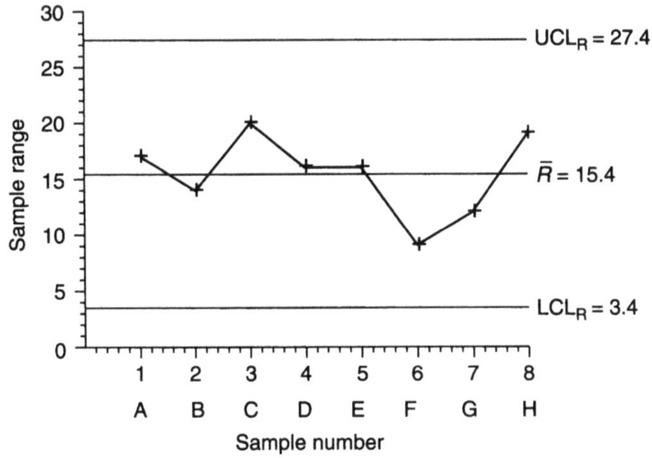

FIGURE 10.3
R-chart for Example 10.2.

and

$$d_3 = 0.80$$

By substituting the above three values into Equation (10.5) and Equation (10.6) we get

$$\text{UCL}_R = \left[1 + \frac{3(0.80)}{(3.08)}\right](15.4) = 27.4$$

and

$$\text{LCL}_R = \left[1 - \frac{3(0.80)}{(3.08)}\right](15.4) = 3.4$$

For $\text{UCL}_R = 27.4$, $\text{LCL}_R = 3.4$, $\overline{R} = 15.4$, and the values of Table 10.3 the *R*-chart is shown in Figure 10.3.

In the figure all Table 10.3 values lie within the control limits, which means that there is no abnormality whatsoever.

10.2.3 The $\overline{X}$ Charts

These charts belong to the family of control charts for variables and are also known as control charts for averages. They are used to indicate the variation in the mean value of the process. Sample means are plotted on the $\overline{X}$ chart.

The distribution of sample means becomes approximately normal when the size of each sample is four or more [8].

The average of the sample means is expressed by

$$\bar{\bar{x}} = \frac{\sum_{i=1}^{m} \bar{x}_i}{m} \tag{10.7}$$

where m is the total number of samples and $\bar{x}_i$ is the mean of sample i, for $i = 1, 2, \ldots, m$.

The upper and lower control limits for the $\overline{X}$ chart are given by

$$\mathrm{UCL}_{xb} = \bar{\bar{x}} + 3s \tag{10.8}$$

and

$$\mathrm{LCL}_{xb} = \bar{\bar{x}} + 3s \tag{10.9}$$

where

UCL_{xb} = upper control limit of the $\overline{X}$ chart
LCL_{xb} = lower control limit of the $\overline{X}$ chart
s = standard deviation of the sample averages

In practice, to simplify calculations, UCL_{xb} and LCL_{xb} are defined as follows [7]:

$$\mathrm{UCL}_{xb} = \bar{\bar{x}} + W\bar{R} \tag{10.10}$$

and

$$\mathrm{LCL}_{xb} = \bar{\bar{x}} - W\bar{R} \tag{10.11}$$

where

$$W = \frac{3}{n^{1/2} d_2} \tag{10.12}$$

where n is the size of the sample.

Example 10.3

Using the Example 10.2 data develop the $\bar{x}$ chart.

The mean $\bar{R}$ of the Example 10.2 data sample ranges is

$$\bar{R} = 15.4$$

TABLE 10.4

Samples A–H Mean Values

	Samples							
	A	B	C	D	E	F	G	H
Sample mean	14.1	15.3	17.2	17.1	18	11.8	18.4	16.1

Using the Table 10.2 data, the means of samples A–H are tabulated in Table 10.4.

By substituting the Table 10.4 values into Equation (10.7) yields

$$\bar{\bar{x}} = 16$$

Using the specified data in Table 10.1 and then the resulting figure and the specified data in Equation (10.12) yields

$$W = \frac{3}{(10)^{1/2}(3.08)} = 0.31$$

By inserting the above three calculated values into Equation (10.10) and Equation (10.11), we get

$$UCL_{xb} = 16 + (0.31)(15.4) = 20.77$$

and

$$LCL_{xb} = 16 - (0.31)(15.4) = 11.23$$

Using the Table 10.4 data and the above calculated values, the $\overline{X}$-chart shown in Figure 10.4 was developed. The chart shows that all Table 10.4 data values are within upper and lower control limits. It means that there is no abnormality whatsoever.

10.2.4 The *c*-Charts

Time to time these charts are also referred to as the control charts for defects per unit. The *c*-chart is used where the subgroup size is *m* inspected unit such as an airplane, a canoe, and a ream of paper. More specifically, the chart is used to control the occurrence of the total number of defects per unit when the subgroup size does not vary or remains constant.

In general, it may be added that the *c*-chart is used where the chances for defect occurrence in an item are high. The *c*-chart mathematics is based on the Poisson distribution. Thus, the mean $\bar{c}$ of the Poisson distribution in the term of *c*-chart is given by

$$\bar{c} = \frac{TNF}{TNI} \tag{10.13}$$

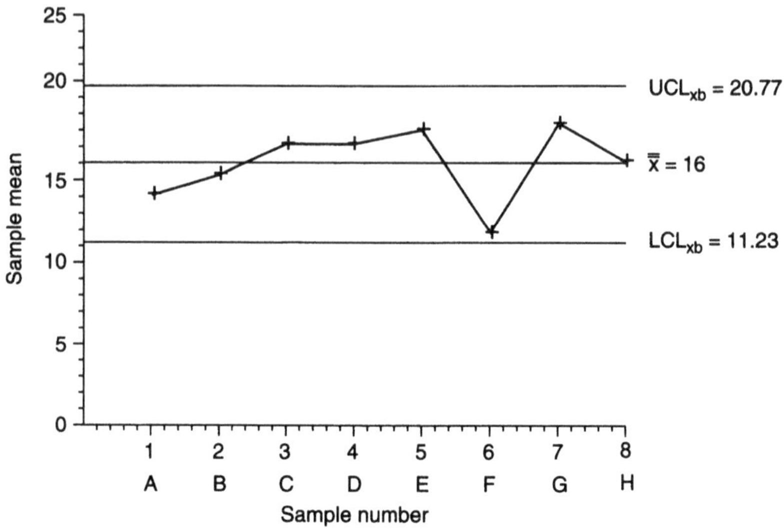

FIGURE 10.4
x-chart for Example 10.2.

where TNF is the total number of defects and TNI is the total number of items.

Thus, the standard deviation is expressed by

$$\sigma_p = \sqrt{\bar{c}} \tag{10.14}$$

where σ_p is the standard deviation of the Poisson distribution.

The upper and lower control limits of the *c*-chart are expressed by [5]

$$UCL_c = \bar{c} + 3\sigma_p \tag{10.15}$$

and

$$LCL_c = \bar{c} - 3\sigma_p \tag{10.16}$$

where UCL_c is the upper control limit of the *c*-chart and LCL_c is the lower control limit of the *c*-chart.

Example 10.4

A total of 12 pumps were inspected for defects. Pumps 1, 2, 3, 4, 5, 6, 7, 8, 9, 10, 11, and 12 contained 15, 10, 7, 13, 10, 12, 18, 8, 10, 6, 11, and 12 defects, respectively. Develop the *c*-chart.

Total number of pumps = 12

The combined number of defects of all pumps is given by

$$= 15+10+7+13+10+12+18+8+10+6+11+12$$
$$= 132$$

By substituting the above two values into Equation (10.13) yields

$$\bar{c} = \frac{132}{12} = 11 \text{defects/pump}$$

Using the above value in Equation (10.14), we get

$$\sigma_p = \sqrt{11} = 3.32$$

Inserting the above calculated values into Equation (10.15) and Equation (10.16) we get

$$UCL_c = 11 + 3(3.32) = 20.95$$

and

$$LCL_c = 11 - 3(3.32) = 1.05$$

Using the above results and the specified data, the Figure 10.5 *c*-chart was developed. The chart depicts that all specified data values of Example 10.4 are well within the upper and lower control limits. It means that there is no abnormality.

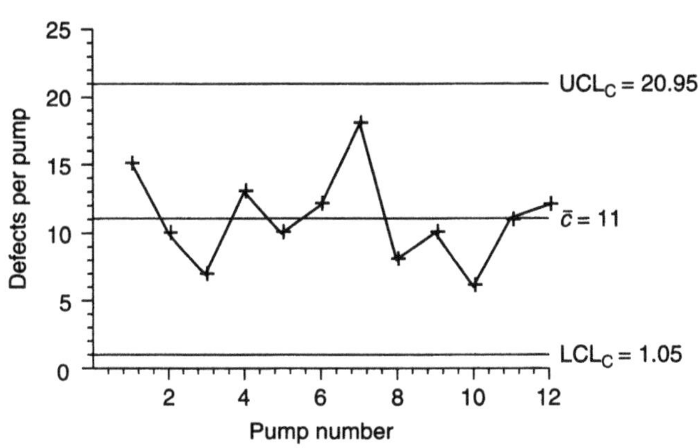

FIGURE 10.5
c-chart for Example 10.4.

10.3 Pareto Diagram

This diagram may simply be described as a simple bar chart that ranks related measures/problems in decreasing frequency of occurrence. The diagram is named after Vilfredo Pareto (1848–1923), an Italian economist and sociologist, who performed a study on the spread of wealth and poverty in Europe in the early 1900s. He was surprised to learn that wealth was concentrated in the hands of around 20% of the people and poverty in the hands of about 80% of the individuals. His findings may simply be called the Law of the "significant few vs. the trivial many."

More specifically, the significant few things will normally make up around 80% of the whole and the trivial many about 20%. Thus, the objective of a Pareto diagram is to highlight the significant aspects of a given problem from the trivial ones. This, in turn, helps the decision maker where to direct improvement efforts.

In quality control work, the Pareto principle was introduced by Juran [9, 10]. He strongly believed that there were always a few kinds of defects in the hardware manufacture that loom large in occurrence frequency and severity. More specifically, around 80% of the scrap is caused by about 20% of the problems.

A Pareto diagram can be constructed by following the steps listed below [8]:

- Determine the approach of classifying the data (i.e., by cause, type of nonconformity, problem, etc.).
- Decide what is to be used to rank the characteristics (i.e., dollars or frequency).
- Obtain all required data (i.e., for an appropriate time interval).
- Summarize the data.
- Rank classifications from largest to smallest.
- Calculate the cumulative percentage, if necessary.
- Construct the Pareto diagram.
- Determine the vital few.

10.4 Quality Function Deployment (QFD)

This is an effective approach used for optimizing the process of developing and manufacturing new products according to customer requirements. QFD was developed in Japan in the late 1960s by two Professors, Mizuno and Akao [11, 12]. It was first introduced into U.S. and Europe in 1983 [2, 11–13].

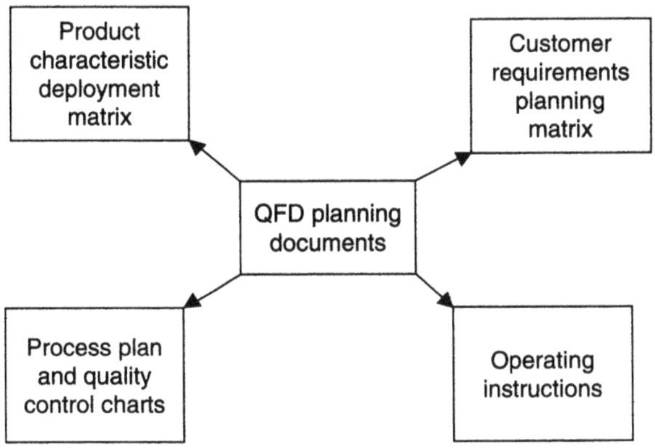

FIGURE 10.6
Important quality function deployment (QFD) planning documents.

In a broader context, QFD may simply be described as a formal process used to translate customer needs into a set of technical requirements. The method uses a set of matrices to relate customer requirements to counterpart characteristics that are expressed as technical specifications and process control requirement. The important QFD planning documents are shown in Figure 10.6.

Process plan and quality control charts identify key process and product parameters along with control points. The customer requirements planning matrix translates the consumer needs into product counterpart characteristics. The product characteristic deployment matrix translates final product counterpart characteristics into critical component characteristics. The operating instructions identify operations that must be completed successfully to achieve critical parameters.

Often, a QFD matrix is called the "House of Quality" because of its resemblance to the structure of a house. The main steps required to build the house of quality are shown in Figure 10.7 [2, 11–13].

The main advantage of the QFD is that it helps to encourage organizations to focus on the process itself rather than focusing on the service or product. Consequently, it reduces engineering design changes and start-up costs and increases customer satisfaction [8].

The key limitation of the QFD is that the exact requirements must be identified in complete detail.

10.5 Scatter Diagram

A scatter diagram is a useful tool to study relationships between two variables. One variable is plotted on the vertical axis and the other on the horizontal axis. Although scatter diagrams cannot prove that one variable causes the

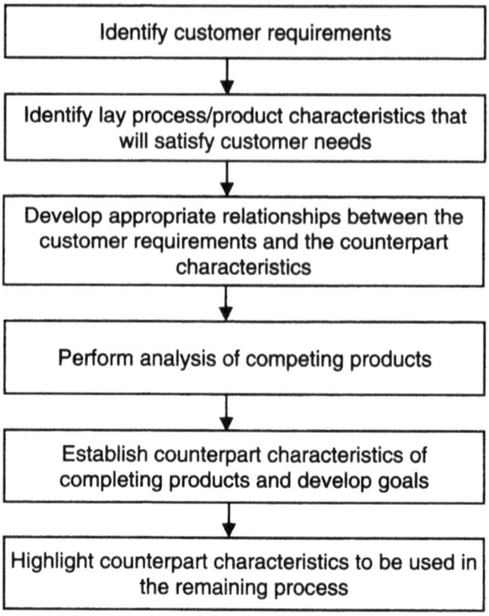

FIGURE 10.7
Steps for building the house of quality.

other, they do indicate the existence of a relationship and its strength. Scatter diagrams can be quite useful in both quality control and problem solving because they display what happens to one variable when another variable is changed. The slope of the diagram indicates the type of relationship between the variables [10].

Usually, scatter diagrams show one of six possible correlations between the variables. These are: strong positive correlation (i.e., the value of variable Y clearly increases as the value of variable X increases), weak positive correlation (i.e., the value of variable Y increases slightly as the value of variable X increases), strong negative correlation (i.e., the value of variable Y clearly decreases as the value of variable X increases), weak negative correlation (i.e., the value of variable Y decreases slightly as the value of variable X increases), complex correlation (i.e., the value of variable Y seems to be related to the value of variable X, but the relationship is not easily determined), and no correlation whatsoever (i.e., there is no demonstrated connection between variables X and Y).

The following four steps are involved in the construction of a scatter diagram [8]:

- Collect two pieces of data (a pair of numbers) on a process, product, or any other item under consideration and develop a summary table of the data.
- Construct a diagram labeling the vertical and horizontal axes.

- Plot the data pairs on the diagram by placing, say a cross, at the intersections of the variables X and Y coordinates for each pair of data.
- Make interpretation of the final scatter diagram for direction and strength.

10.6 Cause-and-Effect Diagram

This was developed by a Japanese named Kaoru Ishikawa in 1943 and basically is a picture made up of symbols and lines designed to denote a meaningful relationship between an effect and its associated causes [8]. The diagram is also known as Ishikawa diagram (i.e., after its originator) or "fishbone" diagram because of its striking resemblance to the skeleton of a fish as shown in Figure 10.8. The rectangle (i.e., the effect) on the right-hand side in the figure denotes the "Fish head" and the left of the rectangle denotes all the possible effect causes which are connected to the central line called the "Fish spine."

From the standpoint of quality, in Figure 10.8, the effect is the quality characteristic that requires improvement and causes 1, 2, 3, 4, 5, and m denote major causes of work methods, people, materials, measurement, equipment, and environment, respectively. Each of these major causes is made up of numerous minor causes identified as subcauses in Figure 10.8.

Past experiences indicate that the cause-and-effect diagram is an extremely useful method for determining the root causes of a problem and generating relevant ideas. There are unlimited opportunities for the application of this

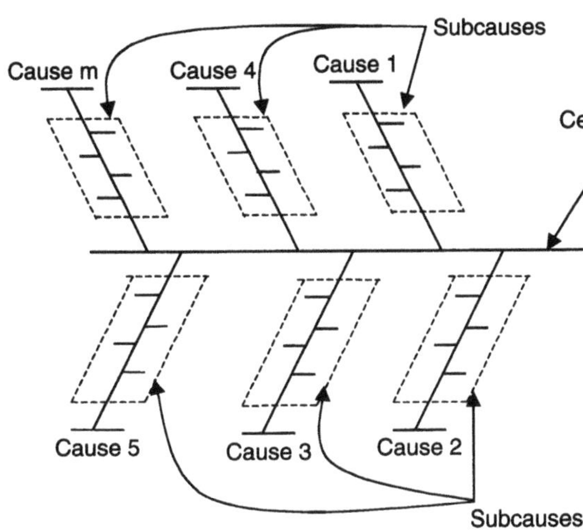

FIGURE 10.8
Cause and effect diagram.

method in areas such as research, manufacturing, office operations, and marketing.

Generally, the steps listed below are followed in developing a cause-and-effect diagram:

- Establish problem statement.
- Brainstorm systematically to identify all possible causes.
- Develop major cause categories by stratifying into natural groupings and process steps.
- Draw the diagram by connecting the causes by carefully following the process steps and fill in the effect or the problem in box on the extreme right hand side of the diagram.
- Refine cause categories by asking questions such as why does this condition exist? And what causes this?

All in all, some of the important benefits of the cause-and-effect diagram are a useful tool to identify root causes, an effective method to generate ideas, and an extremely useful approach for presenting an orderly arrangement of theories.

10.7 Hoshin Kanri

This method was developed in Japan and it may simply be described as a systems approach to the management of change in crucial business processes using a systematic planning, implementation, and review process. Hoshin and Kanri are two Japanese words. The word Hoshin can be divided into two parts: ho (means direction) and shin (means needle). Similarly, the word Kanri can also be divided into two parts: kan (means control or channeling) and ri (means reason or logic). Therefore, both the words taken together basically means management and control of the company's direction needle or focus [14, 15].

Hoshin Kanri was widely used in Japan by the mid-1970s and started to creep into the U.S. in the early 1980s [14, 15]. Factors such as listed below pertain to Hoshin Kanri [14–17].

- Identification of measurable goals.
- Establishment of long and short term organizational goals.
- Identification of the key processes in achieving the above goals or objectives.
- Guidance of team members to reach consensus on concerned performance indicators at process stages.
- Challenging each process level to force company or organization to make appropriate changes to its quality culture.

- Using company or organizational goals as measurable goals for making all involved employees understand the importance of the quality improvement process.

10.8 Design of Experiments (DOE)

This is simply a strategy for planning research which was first introduced by an agricultural scientist, Ronald Fisher, in England in the early 1920s [18]. He developed it to show that how valid experiments could be conducted in the presence of many fluctuating conditions such as rainfall, temperature, and soil condition. Since the 1940s, DOE has been applied successfully in industrial and military applications.

Today, it is widely used by companies to solve serious problems afflicting their operations because the approach provides information regarding the interaction of factors and the way the entire system functions as well as it shows how interconnected factors respond over a wide range of values without testing all possible values directly. Nonetheless, the DOE aim is to determine critical product or process variables and their target values.

There are following three approaches to DOE [8]:

- **Classical.** This is based on the works of Ronald Fisher in the area of agriculture.
- **Shainin.** This was developed by Dorian Shainin and uses numerous problem-solving methods after the product is put in production.
- **Taguchi.** This was developed by Genichi Taguchi and is basically a simplified classical approach with additional engineering design concepts.

All in all, it is recommended that practitioners should first become familiar with these three approaches and then formulate their own methodology. The DOE method is described in detail in Ref. [19].

10.9 Fault Tree Analysis (FTA)

This is a widely used method in industry to evaluate reliability and safety of engineering systems. It can also be used to perform various types of quality-related analysis. The method was developed at the Bell Telephone Laboratories in the early 1960s to evaluate reliability and safety of the Minuteman Launch Control System [7].

The method is described in detail in Chapter 6 and the following example demonstrates its application to quality-related problems.

Example 10.5

Assume that a company manufactures an engineering product. The manufactured product could be of poor quality or defective due to three factors: poor design, poor quality control, or poor manufacturing equipment. Poor product design can result from any of the following three factors:

- Poorly written design specification
- No consideration given to quality
- Design produced in haste

Causes for poor quality control are carelessness, poorly written quality procedures, and poorly trained quality personnel. More specifically, any of these three causes can lead to poor quality control. Using fault tree symbols given in Chapter 6 develop a fault tree for the undesired event: Poor quality or defective product.

A fault tree for the example is shown in Figure 10.9.

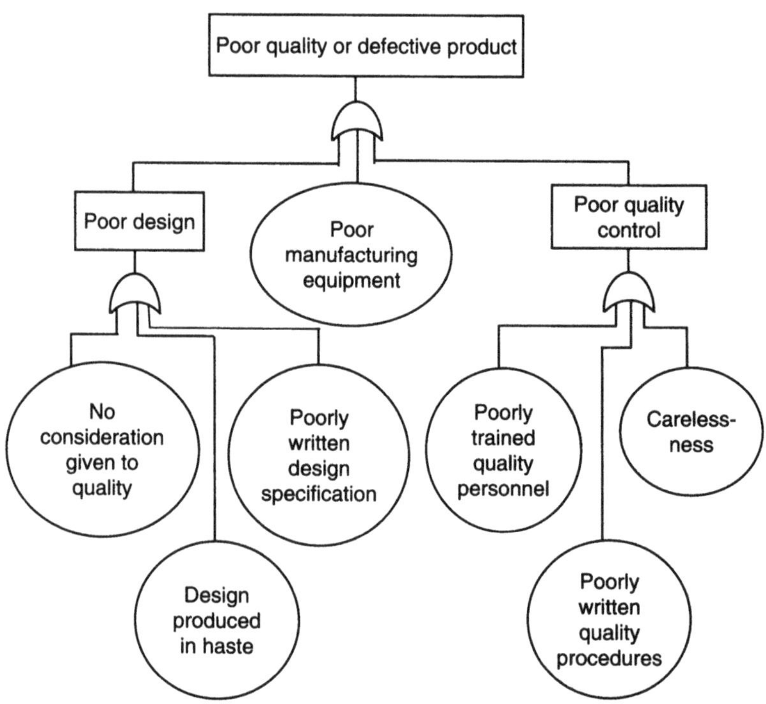

FIGURE 10.9
Fault tree for Example 10.5.

10.10 Failure Modes and Effect Analysis (FMEA)

This method was originally developed in the reliability field and can also be used to perform quality analysis. The method is described in detail in Chapter 6. The following basic seven steps are associated with the method [20–21].

- Define system under consideration and its associated requirements.
- Establish ground rules.
- Describe system and its associated hardware.
- Describe system functional blocks.
- Identify failure modes and their effects.
- Compile a list of critical items by examining the above analysis.
- Document the analysis and take appropriate corrective measures.

10.11 Problems

1. What is a quality control chart?
2. Discuss the following terms with respect to quality control charts:
    - Upper control limit
    - Lower control limit
    - Random samples
3. Discuss the following two types of quality control charts:
    - The *p*-chart
    - The *c*-chart
4. Write down steps for constructing a Pareto diagram.
5. Write down steps for building the "house of quality."
6. What is quality function deployment?
7. Discuss scatter diagram.
8. What are the other names used for cause-and-effect diagram?
9. Discuss the followings two items:
    - Hoshin Kanri
    - Design of experiments
10. A total of ten electric motors were inspected for defects. Motors 1, 2, 3, 4, 5, 6, 7, 8, 9, and 10 contained 8, 10, 15, 16, 11, 5, 20, 19, 12, and 9 defects, respectively. Develop the *c*-chart.

References

1. Ryan, T.P., *Statistical Methods for Quality Improvements*, John Wiley and Sons, New York, 2000.
2. Mears, P., *Quality Improvement Tools and Techniques*, McGraw-Hill, New York, 1995.
3. Juran, J.M., Early SQC: A historical supplement, *Quality Progress*, Vol. 30, No. 9, 1997, pp. 73–81.
4. Rosander, A.C., *Applications of Quality Control in the Service Industries*, Marcel Dekker, New York, 1985.
5. Evans, J.R. and Lindsay, W.M., *The Management and Control of Quality*, West Publishing Company, New York, 1996.
6. Vance, L.C., A bibliography of statistical quality control chart techniques, 1970–1980, *Journal of Quality Technology*, Vol. 15, No. 12, 1983, pp. 225–235.
7. Dhillon, B.S., *Quality Control, Reliability, and Engineering Design*, Marcel Dekker, New York, 1985.
8. Besterfield, *Quality Control*, Prentice Hall, Upper Saddle River, New Jersey, 2001.
9. Juran, J.M., Ed., *Quality Control Handbook*, McGraw-Hill, New York, 1974, pp. 2.16–2.19.
10. Smith, G.M., *Statistical Process Control and Quality Improvement*, Prentice Hall, Upper Saddle River, New Jersey 2001.
11. Mizuno, S. and Akao, Y., Ed., *QFD: the customer-driven approach to quality planning and deployment*, Asian Productivity Organization, Tokyo, 1994.
12. Akao, Y., *Quality Function Deployment: Integrating Customer Requirements into Product Design*, Productivity Press, Cambridge, MA, 1990.
13. Bossert, J.L., *Quality Function Deployment: A Practioner's Approach*, ASQC Quality Press, Milwaukee, Wisconsin, 1991.
14. Bechtell, M.L., *The management compass: steering the corporation using Hoshin planning*, American Management Association, New York, 1995.
15. Cowley, M. and Domb, E., *Beyond Strategic Vision: Effective Corporate Action with Hoshin Planning*, Butterworth-Heinemann, London, 1997.
16. Kanji, G.K. and Asher, M., *100 Methods for Total Quality Management*, Sage Publications, London, 1996.
17. Akao, Y., *Hoshin Kanri: Policy Deployment for Successful TQM*, Productivity Press, Cambridge, Massachusetts, 1991.
18. Kim, J.S. and Kalb, J.W., Design of experiments: an overview and application example, *Medical Device and Diagnostic Industry (MDDI) Magazine*, March 1996, pp. 78–83.
19. Montgomery, D.C., *Design and Analysis of Experiments*, John Wiley and Sons, New York, 1990.
20. Arnzen, H.E., Failure Modes and Effect Analysis: A Powerful Engineering tool for component and system optimization, *Annuals of the Reliability and Maintainability Conference*, 1966, pp. 355–371.
21. Jordan, W.E., Failure modes, effects and criticality analysis, *Proceedings of the Annual Reliability and Maintainability Symposium*, 1972, pp. 30–37.

11

Quality Management and Costing

11.1 Introduction

Management plays an instrumental role in the effective performance of quality-related activities in an organization. A well-managed quality organization can be useful in many ways including better communication, less number of conflicts in responsibilities and activities, better customer satisfaction, and lower cost. More simply, a well-managed quality organization is essential in producing an acceptable quality item at competitive prices. Otherwise, in today's global economy, a company with poorly managed quality function may not be in business for a very long time.

The cost of quality plays a pivotal role in management decision making. In business terms, it provides the economic common denominator through which individuals involved with management and quality can interact effectively. Similarly, in monetary terms, it is extremely useful in measuring the effectiveness of the quality department, quality planning, and so on. Moreover, the cost of quality could be a key factor in the survival of companies. For example, in the fiscal year 1977 for Firestone the cost of replacing 7.5 million poor quality tires in a recall case was around $135 million after taxes, i.e., greater than the company's net income for the year [1].

This chapter presents various important aspects of quality management and costing.

11.2 Upper and Middle Management Quality-Related Roles

The active leadership by upper management is of all the ingredients for successfully achieving superiority in the quality area. Some of the upper management roles with respect to quality are presented in Figure 11.1 [2].

The quality strategy developed by the upper management is executed by people such as middle managers, supervisors, professional specialists, and the workforce. Some of the roles played by middle managers, supervisors,

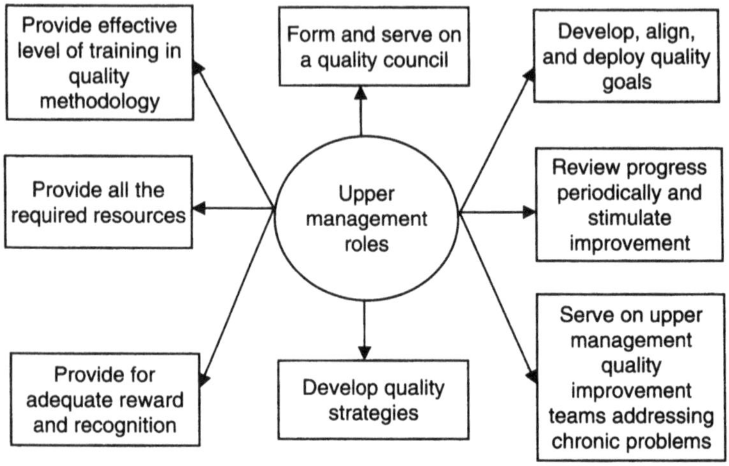

FIGURE 11.1
Important upper management quality-related roles.

and specialists are nominating quality problems for finding solutions, serving on task forces as the need arises, to assist the quality council in developing quality strategy elements, serving as members of quality teams, serving as leaders of various types of quality teams, identifying suppliers and customers and periodically meeting with them to determine and address their pressing requirements, and leading the quality activities within their own area of responsibility by demonstrating a personal commitment and encouraging their employees [2].

11.3 Quality Control Engineering Functions and Quality-Related Responsibilities Among Various Organizational Groups

Quality control engineering performs numerous functions. They may be grouped under seven distinct categories as shown in Figure 11.2 [3]. These are process quality control, design quality control, product evaluation quality control, inventory evaluation quality control, special studies, incoming material quality control, and general quality control.

There are many areas of responsibility with respect to quality and there are many organizational groups who are responsible for them. Most of these areas of responsibility (along with their corresponding responsible organizational groups), are to determine the needs of customers (marketing), plan the quality system (general management/manager, quality control), establish product design specifications (engineering), produce products according to design

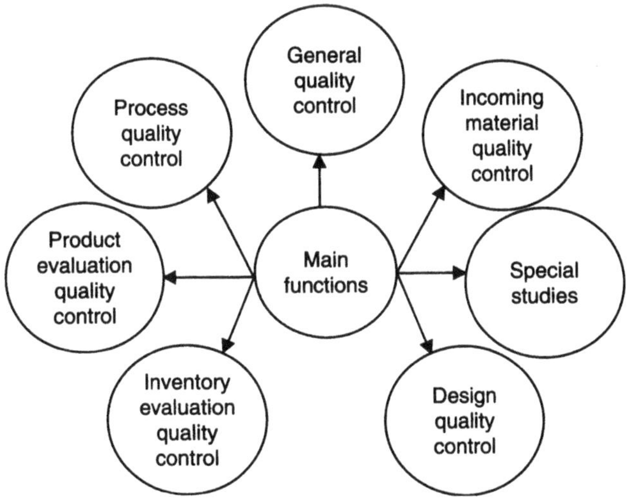

FIGURE 11.2
Main functions of the quality control engineering.

specifications (shop operations), collect complaint data (marketing), develop manufacturing process design (manufacturing engineering), develop quality level for business (general management/manager), analyze complaint data (quality control), determine process capabilities (quality control), compile quality costs (finance), qualify suppliers on quality (materials), analyze quality costs (quality control), design test and inspection equipment (quality control), final product inspection (quality control), plan inspection and test procedures (quality control), in-process quality measurements (quality control), in-process quality audit (quality control), feedback quality information (quality control), and obtain corrective action (quality control) [4, 5].

11.4 Steps for Planning the Quality Control Organizational Structure and Quality Control Organizational Methods

Over the years many professionals have outlined various steps for planning the quality control organizational structure. The following steps are considered the most useful to achieve this goal effectively [5, 6]:

- Define the company/enterprise goals for which the quality organization is being formed.
- Develop objectives for implementing the enterprise goals.
- Identify basic work activities for achieving the set objectives.
- Group work activities into basic functions.

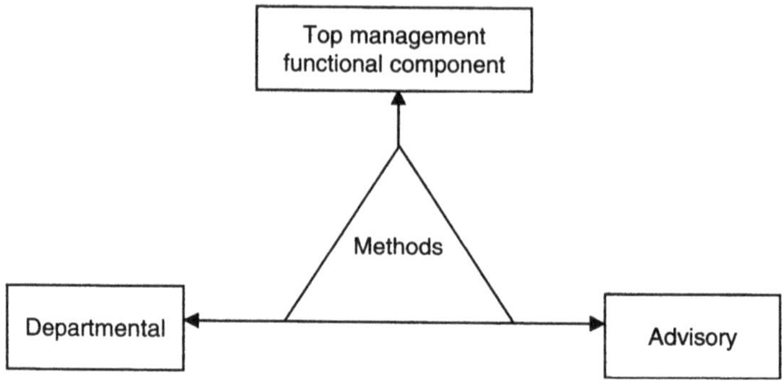

FIGURE 11.3
Common quality control organizational methods.

- Combine the functions into job packages.
- Consolidate job packages into an organizational unit for satisfying enterprise needs.
- Locate the unit in that part of the enterprise organizational set up or structure where its (enterprise) goals can be achieved most effectively.

There are many quality control organizational methods. The frequently used methods are shown in Figure 11.3 [7].

In the case of "top management functional component" method, the quality control is made a functional element of upper management. More specifically, the position of the quality chief is quite similar to that of the controller and he/she reports directly to the president. The main advantage of this method is that the quality chief can exercise his/her authority to place quality-related responsibilities at appropriate places within the organization.

In the case of "departmental" method a separate department is created to perform the quality function. The head of the department reports to the general manager, who in turn reports to the president. The main advantage of this method is that it provides a definite responsibility for quality to the department chief. In turn, this gives good incentives to the individual to produce effective results.

In the case of "advisory" method quality personnel such as inspectors, inspector supervisors, etc. act as "process advisors." More specifically, these individuals do not exercise authority but simply act as advisors to production and engineering departments. They report to a general manager, who in turn reports directly to the president. The main advantage of this method is that it does not require substantial changes and reduces the probability of having differences between production and inspection people. In contrast, its main drawbacks are the lack of authority and no definite responsibilities for quality personnel.

11.5 Quality Manager Attributes, Functions, and Reasons for Failure

Just like in the case of any other manager, the quality manager must also possess certain attributes to perform his/her task effectively. Most of these attributes are good communication skill, broad knowledge of engineering, research, and production, negotiating skill, ability to plan, organize, delegate, and sell effectively, competence in statistical analysis, ability to listen and to motivate, ability to think effectively and to utilize time effectively, creativeness and fairness, enthusiasm, honesty and compassion, impartiality and decision-making ability, patience and persistence, knowledge of finance and marketing functions, performance-mindedness, and good health and strength of character [6–8].

The quality manager performs various types of functions. Some of the important ones are shown in Figure 11.4 [9].

There are numerous reasons for the failure of quality managers to perform their job effectively. Some of these are as follows [6]:

- Failure to communicate effectively with right individuals.
- Overlooking to emphasize fitness for use; instead emphasizing conformance to specifications.
- Placing emphasis on problems for which there are already satisfactory solutions.
- Failure to diagnose the problem first, instead emphasizing special skills as solution to all problems.
- Failure to secure appropriate inputs from all desirable/important individuals because of functioning as loners.

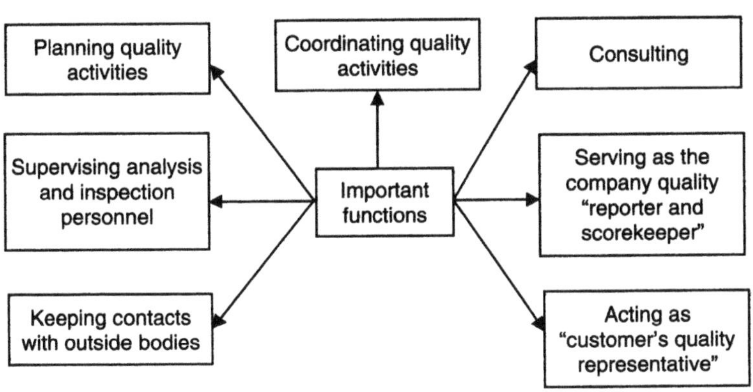

FIGURE 11.4
Important functions of a quality manager.

11.6 Quality Control Manual and Quality Auditing

Quality control manual is a very important document and it contains various types of information on the quality control program of an organization. Usually, the manual contains information on items such as organizational charts, responsibilities, quality policies and procedures, inspection procedures, quality costs, marketing, manpower, statistical methodology, defect prevention, measuring equipment, and vendor quality control procedure [5].

There are many benefits of having a quality control manual including useful for making various quality related decisions, a reference document, useful for the continuity of quality control organization operations despite the manpower turnover, and it can be used as a textbook when training quality manpower.

Auditing is an important element of quality control. There are various reasons for its performance including to determine whether operators perform their tasks according to specified quality plans, to determine whether the end product meets the important quality specifications, and to determine whether equipment and machinery are operating according to expectations [10].

Some of the useful auditing guidelines are as follows [11]:

- Use checklists to conduct audits.
- Choose an unbiased person to perform audits.
- Perform audits without prior announcements.
- Maintain audit schedules.
- Avoid conducting audits with the intention of catching someone performing his/her tasks incorrectly.
- Ensure that no individual performs quality audits for more than six months at a time.
- Do not make deals with people when auditing.
- Ensure that all audit results are recorded effectively.
- Audit all the work shifts involved.
- Distribute the auditing results to all concerned individuals.
- Take all necessary follow-up actions.

11.7 Procurement Quality Control

Usually, product manufacturers procure many parts and materials from outside suppliers. In the U.S. companies, over 50% of revenues from product sale is spent on purchasing parts, materials, supplies, etc. from other companies [11].

Moreover, the quality of incoming parts and materials has become more and more critical with the increase in automation. Needless to say, in order to produce good quality products, the quality of procured parts and materials used has to be monitored properly.

Nonetheless, some of the specific reasons for having procurement quality assurance are to determine whether the technical and contractual requirements are adequately specified in the request for proposal document, to help the suppliers of procured parts/materials understand the specified requirements, to rate vendor conformance, to rate vendor performance, and to determine whether the requirements are being satisfied effectively [12].

11.7.1 Useful Guidelines for Controlling Incoming Parts/Materials

Some of these guidelines are as follows [13]:

- Develop close relationships with suppliers.
- Develop compatible quality measurement procedures among suppliers and users.
- Ensure the availability of satisfactory storage facilities.
- Audit and survey part/material suppliers.
- Use acceptance sampling tables.
- Ensure the availability of effective material handling equipment and services.
- Establish a system for disposing immediately nonconforming materials and parts.
- Review, periodically, the effectiveness of inspecting incoming parts and materials.
- Ensure that measuring equipment and gauges used in the quality work are maintained in good condition.
- Ensure the availability of satisfactory receiving inspection and test facilities.
- Use statistical methods to analyze data on incoming materials.
- Provide adequate training to the inspectors of incoming materials and parts.
- Make quality procurement and other necessary information available to vendors.

11.7.2 Incoming Material Inspection

Usually, newly purchased items are inspected after their arrival at the buyer's facility. Some of the reasons for conducting receiving inspection are high probability for having defective items in the order, the purchased items may be unsafe to workers, and the defective shipment can degrade the efficiency of manufacturing and outgoing quality.

The inspection of incoming materials may take various different forms including the inspection of all received (purchased) items (i.e., 100% inspection) with respect to the conformance to the required specifications, taking a sample of the received lot to determine if the lot conforms to the required specifications and then making accept or reject decision, and checking the received items to determine if they are the same as ordered [14].

11.7.3 Formulas for Determining Accuracy and Waste of Inspectors and Vendor Quality Rating

There is a possibility that inspectors can reject good items and accept bad ones. The check inspectors can be used to review the output of regular inspectors to minimize the occurrence of this scenario. More specifically, the check inspectors re-examine the procedure followed by the regular inspectors as well as their total output (i.e., all accepted and rejected items).

Two formulas to estimate accuracy and waste of regular inspectors are presented below [9–15].

11.7.3.1 Regular Inspector Accuracy Estimation Formula

The percent of defects correctly identified by the regular inspector is expressed by

$$PDC = \left[\frac{N - M}{N - M + \theta}\right](100) \tag{11.1}$$

where

PDC = percent of defects correctly identified by the regular inspector
N = number of defective items found by the regular inspector
M = number of items without defects rejected by the regular inspector as revealed by the check inspector.
θ = number of defective items missed by the regular inspector as revealed by the check inspector.

11.7.3.2 Regular Inspector Waste Estimation Formula

The percent of good items rejected by the regular inspector is expressed by

$$PGIR = \frac{M(100)}{TI - (N - M + \theta)} \tag{11.2}$$

where PGIR is the percent of good items rejected by the regular inspector and TI is the total number of items rejected.

Example 11.1

Assume that a regular inspector inspected a certain number of items in a lot and found 30 defective. Subsequently, a check inspector re-examined all the items (i.e., good and defective) and found four items without defects rejected by the regular inspector and five defective items missed by the regular inspector. Calculate the percent of defects correctly found by the regular inspector.

Using the given data in Equation (11.1), we get

$$PDC = \left[\frac{30-4}{30-4+5}\right](100)$$
$$= 83.87\%$$

It means that 83.87% of the defects were correctly found by the regular inspector.

11.7.3.3 Vendor Quality Rating Formula

Vendor quality rating is necessary to ensure adequate quality of purchased items. Usually, the vendor performance is measured by considering factors such as procured items' cost, delivery, and quality. Frequently, weights assigned to cost, delivery, and quality are 40, 30, and 30%, respectively [10]. Nonetheless, vendor quality rating index is defined by

$$VQRI = \frac{LAB(100)}{LRB} \tag{11.3}$$

where

VQRI = vendor quality rating index
LAB = total number of lots accepted by the buyer
LRB = total number of lots received by the buyer (i.e., all lots whether accepted or rejected)

Equation (11.3) does not take into consideration the cost and delivery factors.

Example 11.2

Assume that an engineering product manufacturer received a total of 40 equal shipments of a certain component from a vendor over a period of one year. Two of these shipments were rejected by the incoming material inspection group. Calculate the value of the quality rating index.

Thus, the total number of lots accepted by the manufacturer is given by

$$LAB = (\text{Total no. of lots received}) - (\text{lots rejected})$$
$$= 40 - 2$$
$$= 38$$

Using the above calculated value and the given data in Equation (11.3) yields

$$VQRI = \frac{38}{40}(100)$$
$$= 95\%$$

Thus, the value of the quality rating index is 95%.

11.8 Quality Costs

The concept of "quality costs" has emerged since the 1950s. It may mean different things to different people. For example, some may equate quality costs with the costs of attaining quality and the others may equate them with the extra costs incurred because of poor quality [2].

In the early 1970s, the cost of quality was around 10% of the sales income in many U.S. manufacturing organizations [16].

11.8.1 Classifications of Quality Costs

Quality costs may be classified under five distinct categories [11, 17]:

- Costs of internal failures
- Costs of external failures
- Administrative costs
- Prevention costs
- Appraisal and detection costs

Each of the above classifications is discussed below separately.

11.8.1.1 Costs of Internal Failures

These costs occur after the delivery of the product to the buyer and are associated with items such as components, materials, and products that fail to meet quality requirements. More specifically, the internal failure costs are associated with things such as scrap, in-house components and materials failures, rework, re-inspection and retest, redesign, and failure analysis.

11.8.1.2 Costs of External Failures

These costs occur after the delivery of the product to the buyers. More specifically, costs due to faulty products shipped to the buyers. These costs are associated with items such as failure analysis, repair, investigation of customer complaints, replacement of defective items, liability, warranty charges, and customer defections.

11.8.1.3 Administrative Costs

These costs are associated with administrative-related activities including reviewing contracts, clerical, preparing proposals, forecasting, management, preparing budgets, and performing data analysis.

11.8.1.4 Prevention Costs

These costs are associated with actions taken to prevent the production of defective components, products, and materials. These actions include evaluating suppliers, reviewing designs, coordinating plans and programs, training personnel, calibrating and certifying inspection and test devices and instruments, receiving inspection, collecting quality-related data, and implementing and maintaining sampling plans.

11.8.1.5 Appraisal and Detection Costs

These costs are associated with appraisal and detection actions. Three principal components of such costs are cost of testing, cost of auditing, and cost of inspection (i.e., in-process, source, shipping, receiving, etc.).

11.9 Quality Cost Indexes

Often various quality cost indexes are used by organizations to monitor their performance. The values of these indexes are plotted periodically and their overall trends are monitored. This section presents three such indexes [4, 10, 17, 18].

11.9.1 Index I

This index is defined by

$$\alpha = \frac{\text{TQC}}{\text{DLC}}(100) \tag{11.4}$$

where

α = quality cost index expressed as a percentage
TQC = total quality cost
DLC = direct labour cost

It is to be noted that this index does not provide management with that useful information for decision making and problem diagnosis [4]. Often, the index is used to eliminate inflation effects.

11.9.2 Index II

This index is defined by

$$\theta = \frac{TQC}{TS}(100) \tag{11.5}$$

where θ is the quality cost index expressed as a percentage and *TS* is the total sales.

11.9.3 Index III

This index is defined by

$$\gamma = \left[\frac{TQC(100)}{VO}\right] + 100 \tag{11.6}$$

where γ is the quality cost index and *VO* is the value of output.

The values of γ may be interpreted as follows [4]:

- $\gamma = 100$ (It means that there is no defective output).
- $\gamma = 105$ (It can readily be achieved in a real-life environment).
- $\gamma = 110$–130 (This occurs in organizations where the quality costs are ignored).

11.10 Problems

1. List at least eight important upper management quality-related roles.
2. Discuss at least five main functions of the quality control engineering.
3. Write down the steps for planning the quality control organizational structure.
4. Describe the three commonly used quality control organizational methods.
5. What are the important functions of a quality manager?
6. List at least five key reasons for the failure of quality managers to perform their job effectively.
7. What are the benefits of having a quality control manual?
8. List at least ten useful guidelines for controlling incoming parts/materials.

9. Discuss the following three classifications of quality costs:
    - Costs of internal failures
    - Costs of external failures
    - Prevention costs
10. A regular inspector inspected a certain number of items in a lot and discovered 40 defective. Subsequently, a check inspector re-examined all the items (i.e., good plus defective) and found three items without defects rejected by the regular inspector and four defective items missed by the regular inspector. Calculate the percent of defects correctly found by the regular inspector.

References

1. Reece, R.M., Quality control as an inflation fighter, *Quality Progress*, August 1980, pp. 24–25.
2. Gryna, F.M., *Quality Planning and Analysis*, McGraw-Hill New York, 2001.
3. Charbonneau, H.C. and Webster, G.L., *Industrial Quality Control*, Prentice Hall, Englewood Cliffs, NJ, 1978.
4. Evans, J.R. and Lindsay, W.A., *The Management and Control of Quality*, West Publishing Company, New York, 1989.
5. Feigenbaum, A.V., *Total Quality Control*, McGraw-Hill, New York, 1983.
6. Dhillon, B.S., *Quality Control, Reliability, and Engineering Design*, Marcel Dekker, New York, 1985.
7. Enrick, N.L., *Quality Control and Reliability*, Industrial Press, New York, 1972.
8. Dhillon, B.S., *Engineering and Technology Management: Tools and Applications*, Artech House, Norwood, Massachusetts, 2002.
9. Juran, J.M., Gryna, F.M. and Bingham, R.S., *Quality Control Handbook*, McGraw-Hill, New York, 1974.
10. Lester, R.H., Enrick, N.L., and Mottely, H.E., *Quality Control for Profit*, Industrial Press, New York, 1977.
11. Hayes, G.E. and Romig, H.G., *Modern Quality Control*, Collier-Macmillan, London, 1977.
12. Gage, W.G., Procurement quality planning and control, *Proceedings of the Annual American Society for Quality Control Conference*, 1978, pp. 158–161.
13. Feigenbaum, A.V., *Total Quality Control*, McGraw-Hill, New York, 1983.
14. Juran, J.M. and Gryna, F.M., *Quality Planning and Analysis*, McGraw-Hill, New York, 1980.
15. Juran, J.M., Inspectors' errors in quality control, *Mechanical Engineering*, Vol. 59, No. 10, 1935, pp. 643–644.
16. Editorial, *Quality Management and Engineering Magazine*, Vol. 12, No. 3, 1972, pp. 11.
17. Carter, C.L., *The Control and Assurance of Quality, Reliability, and Safety*, Published by C.L. Carter and Associates, Richardson, Texas, 1978.
18. Sullivan, E. and Owens, D.A., Catching a glimpse of quality costs today, *Quality Progress*, Vol. 16, No. 12, 1983, pp. 21–24.

12

Introduction to Safety

12.1 Need for Safety

The desire to be safe and secure has always been an important concern to humans. For example, early humans took necessary precautions to guard against natural hazards around them. Moreover, in 2000 B.C., Hammurabi, an ancient Babylonian ruler, developed a code known as Code of Hammurabi. This code included clauses on items such as monetary damages against people who caused injury to others and allowable fees for physicians [1, 2].

Today, safety has become a critical issue because every year a very large number of people die and get seriously injured due to workplace and other accidents. For example, in 1996 in the U.S. states according to the National Safety Council (NSC), there were 93,400 deaths and a vast number of disabling injuries due to accidents [3]. The total cost of these accidents was estimated to be around $121 billion. Some of the other factors that are also playing an instrumental role in demanding the need for better safety are government regulations, public pressures, and increasing number of law suits.

12.2 Safety-Related Facts and Figures

Some of the facts and figures directly or indirectly concerned with safety are as follows:

- In 2000, there were around 97,300 unintentional injury deaths in the U.S. Their cost was estimated to be around $512.4 billion [4].
- In the European Union (EU), each year around 5,500 people die due to work-related accidents [5].
- In the U.S., in a typical year around 35 million work hours are lost due to accidents [6].
- In 2000, there were 3.8 deaths per 100,000 workers in the U.S. [7].

- In the European Union (EU), each year over 4.5 million accidents result in more than three days' absence from work [5]. This translates into a total loss of around 146 million working days [5].
- In 1995, the total cost of work-related accidents in the U.S. was estimated to be around $75 billion [8].
- In 2000, a total of 5200 persons died due to work-related accidents in the U.S. [4].
- During the period 1960–2000, the work-related accidental deaths in the U.S. dropped by 60% (i.e., 13,800 in 1960 to 5200 in 2000) [7, 9].
- In the 1990s, the cost of accidents per worker per year in the U.S. was around $420 [1].
- In 2000, the total cost of work-related injuries in the U.S. was around $131.2 billion [7].
- In 1980, American employers spent around $22 billion to insure or self-insure against job-related injuries [10].
- In 1997, three workers in the U.S. were awarded $5.8 million after they sued a computer equipment manufacturing company for musculoskeletal disorders (MSDs) [11]. The workers believed that these disorders were caused by keyboard entry activities.

12.3 Engineers and Safety

In the modern times, the problems with the safety of engineering products may be traced back to railroads. For example, the very day Stephenson's first railroad line was dedicated, a fatal railroad accident occurred (i.e., a prominent English legislator was killed) [11]. A year later, the boiler of the first locomotive built in the U.S. exploded and killed one man and badly injured a number of fuel servers [11, 12].

Today, engineering products have become very complex and sophisticated. Their safety is a challenging issue, because of competition and other factors engineers are pressured to complete new designs rapidly and at lower costs. Past experiences indicate that this, in turn, usually results in more design deficiencies, errors, and causes of accidents. Design deficiencies can cause or contribute to accidents. The design deficiency may result because a designer/design [11]:

- Is unfinished, incorrect, or confusing.
- Overlooked to warn adequately of a potential hazard.
- Creates an unsafe characteristic of an item.
- Overlooked to foresee an unexpected application of an item or its all potential consequences.

- Overlooked to reduce or eliminate the occurrence of human error.
- Violates usual capabilities/tendencies of potential users.
- Relies on product users to avoid an accident.
- Places an unreasonable level of stress on operators.
- Does not adequately determine or consider the failure, error, action, or omission consequences.
- Overlooked to provide a satisfactory level of protection in a user's/worker's personal protective equipment.
- Overlooked to prescribe a proper operational procedure in situations where a hazard might exists.
- Creates an arrangement of operating controls and other devices that substantially increases reaction time in emergency situations or is conducive to errors.
- Incorporates poor warning mechanisms instead of providing a rather safe design to eradicate hazards.

12.4 Product Hazard Classifications and Common Mechanical Injuries

There are many product-related hazards. These may be classified under six distinct categories as shown in Figure 12.1 [13].

The electrical hazards have two principal components: electrocution hazard and shock hazard. The major electrical hazard to product or property stem from electrical faults, frequently referred to as short circuits. The energy hazards may be grouped under two categories: kinetic energy and potential energy. The kinetic energy hazards pertain to parts that have energy because

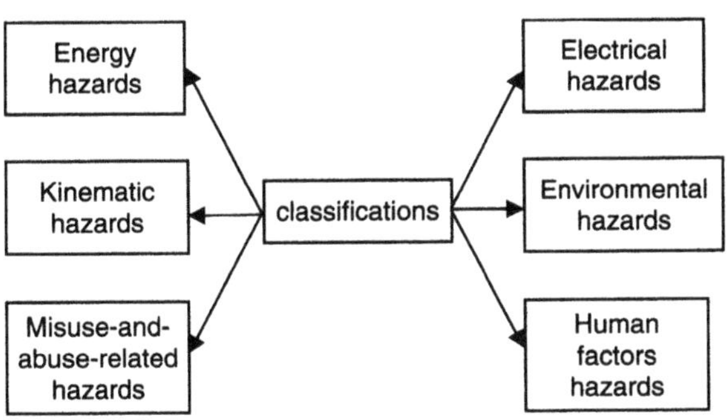

FIGURE.12.1
Product-related hazard classification.

of their motion. Some examples of these parts are flywheels, loom shuttles, and fan blades. Any object that interferes with their motion can experience substantial damage. The potential energy hazards pertain to parts that store energy. Such parts include electronic capacitors, springs, counterbalancing weights, and compressed-gas receivers. During the servicing of the equipment such hazards are important because stored energy can cause serious injury when released suddenly.

The environmental hazards may be grouped under two classifications: internal and external. The internal hazards are associated with the changes in the surrounding environment that lead to internally damaged product. These hazards can be minimized or eliminated altogether by carefully considering factors such as vibrations, atmospheric contaminants, extremes of temperatures, ambient noise levels, electromagnetic radiation, and illumination level during the design phase. The external hazards are the hazards posed by the product under consideration during its life span. These hazards include maintenance hazards, disposal hazards, and services-life operation hazards.

The kinematic hazards are associated with situations where parts come together while moving and lead to possible pinching, crushing, or cutting of any object caught between them. Misuse-and abuse-related hazards are concerned with the product usage by people. Past experiences indicate that product misuse can result in serious injuries. Product abuse can also lead to hazardous situations or injuries and some of the causes for the abuse are poor operating practices and lack of proper maintenance.

The human factor hazards are associated with poor design in regard to people. More specifically, to their intelligence, education, computational ability, height, weight, length of reach, physical strength, visual acuity, visual angle, etc.

In the industrial sector, humans interact with various types of equipment to carry out tasks such as cutting, chipping, drilling, stamping, stitching, abrading, shaping, and punching. There are various types of injuries that can result in performing such tasks. Some of the common ones are shown in Figure 12.2 [1].

The puncturing-related injuries occur when an object penetrates straight into an individual's body and pulls straight out. In the industrial setting, usually, these types of injuries pertain to punching machines because they have sharp tools. The cutting and tearing-related injuries occur when an individual's body part comes in contact with a sharp edge. The severity of a tear or a cut depends upon the degree of damage to items such as muscles, skin, arteries, and veins.

The breaking-related injuries are normally associated with machines used to deform various types of engineering materials. Often, a break in a bone is called a fracture. In turn, fracture is classified into many categories including simple, incomplete, complete, comminuted, transverse, oblique, and compound. The crushing-related injuries occur where a body part is caught between two hard surfaces moving progressively together and crushing any object that comes between them.

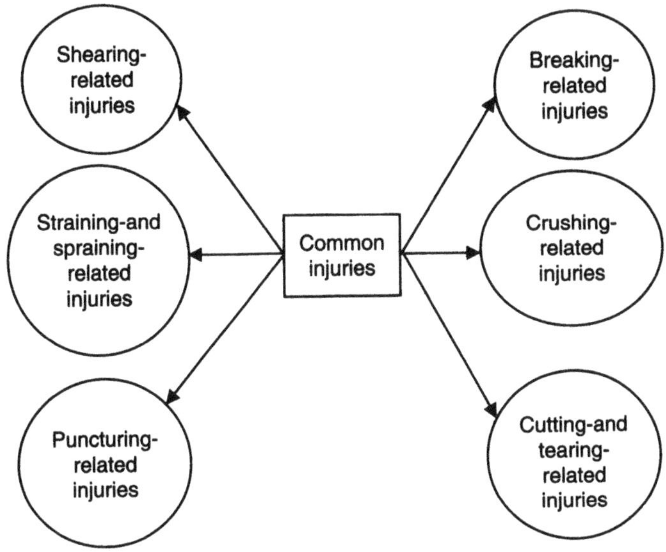

FIGURE 12.2
Common mechanical injuries.

The shearing-related injuries pertain to shearing processes. In manufacturing, power-driven shears are widely used to carry out various types of tasks including severing paper, metal, elastomers, and plastic. In the past, in using such machines tragedies such as amputation of fingers/hands have occurred. In the industrial environment, there are numerous opportunities associated with the use of machines or other tasks, for the occurrence of straining-and spraining-related injuries, e.g., straining of muscles or spraining of ligaments.

12.5 Statute, Common, Administrative, and Liability Laws and Product Liability

Statute laws are promulgated by the highest governing authority or royal edict in any jurisdiction. The history of the statute laws may be traced back to the Code of Hammurabi (circa 2000 B.C.). Hammurabi, an ancient Babylonian ruler, developed this code which contained clauses on areas such as injuries, monetary damages assessed against those who caused injury to others, and allowable fees for physicians [1, 2]. Nonetheless, under the early statute laws a person, who for any reason, caused injury to others was treated as a criminal [11].

Common laws were first used in the UK and subsequently in other English-speaking countries with certain regional variations. They are not established by statutes but follow precedents set forth by previous judicial decisions. Many safety-related cases fall under this category of laws.

Administrative laws are developed by an executive in prescribing the criteria under which any desired control/statute will be executed. Time to time, various safety-related cases also fall under this category of laws.

Liability laws are basically intended to reduce defences of employers with respect to liability for accidents by eradicating altogether the common law concept of assumption of risk. In 1885, Alabama was the first U.S. state to pass an employers' liability law. In 1970, the Occupational Safety and Health Act (OSHA) was passed by the U.S. Congress. Over the years, this Act has helped to improve the safety of workplaces quite dramatically.

Product liability is an important factor in safety. Until 1960, in the U.S., manufacturers were not liable unless they manufactured flagrantly dangerous items or products. However, today manufacturers are increasingly being sued by the product users, misusers, and even by abusers.

Some of the landmark cases that have established the basis for product liability suits in the U.S. are MacPherson vs. Buick Motor Company (this case has established the concept of negligent manufacture), Greenman vs. Yuba Products, Inc., (this case has established the concept of strict liability in tort and negligent design), and Henningson vs. Bloomfield Motors, Inc., (this case has established the concept of breach of warranty) [1].

Past experiences indicate that around 60% of the liability cases involved failure to provide adequate danger warning labels on manufactured products. Nonetheless, some of the common causes of product liability exposure are shown in Figure 12.3 [14].

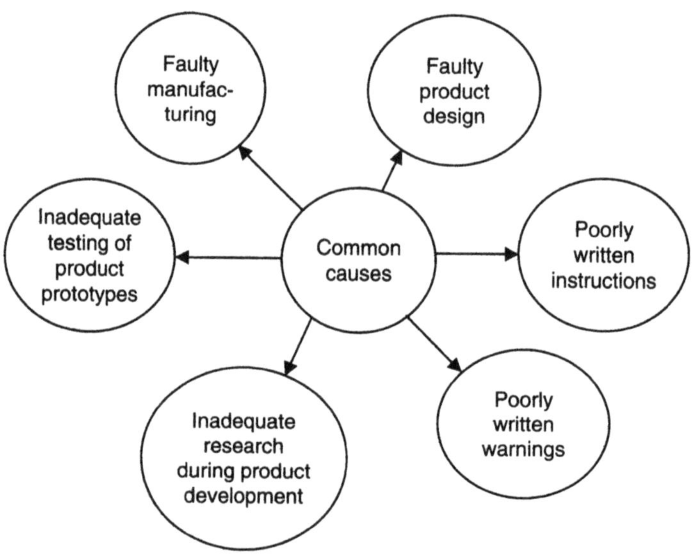

FIGURE 12.3
Common causes of product liability exposure.

12.6 Workers' Compensation

Due to various government legislations workers' compensation has become an important factor in workplace safety. The workers' compensation law was first passed in Prussia in 1838 to protect railroad workers [11]. However, in the U.S., it was only in 1908 when the first workers' compensation law was passed to protect federal government employees performing various types of hazardous tasks. In today's work environment, in general, all American workers are well protected by compensation laws. Nonetheless, in 1983, a study of Workers' Compensation Laws was performed by the U.S. Chamber of Commerce.

It concluded their seven underlying objectives: (i) to encourage frank and fair study of causes of accidents, (ii) to provide an appropriate approach to reduce the degree of personal injury litigation in courts, (iii) to provide fast and reasonable income and medical benefits to victims of work-accidents or income benefits to dependents of victims, irrespective of faults, (iv) to minimize, as much as possible, human suffering and preventable accidents, (v) to eliminate time-consuming and costly trials and appeals, (vi) to free charities from finance-related burdens created by uncompensated workplace accidents, (vii) to maximize employer involvement in safety and rehabilitation through the use of an experience-rating mechanism [11, 14]. The subject of workers' compensation is discussed in detail in Refs. [1, 11, 15–17].

12.7 Problems

1. List at least ten safety-related facts and figures.
2. What are the common causes of product liability exposure?
3. Discuss design deficiency factors with respect to designer.
4. Discuss six classifications of product-related hazards.
5. Discuss the following terms:
    - Potential energy
    - Kinetic energy
    - Internal environmental hazards
    - External environmental hazards
6. Discuss common mechanical injuries.
7. Write an essay on workers' compensation.
8. Discuss product liability.

9. Discuss the following:
   - Statute law
   - Common law
10. Discuss the need for safety.

References

1. Goetsch, D.L., *Occupational Safety and Health,* Prentice Hall, Englewood Cliffs, NJ, 1996.
2. Ladon, J., Ed, *Introduction to Occupational Health and Safety,* National Safety Council (NSC), Chicago, IL, 1986.
3. Accidental Facts, Report, National Safety Council, Chicago, IL, 1996.
4. Report on Injuries in America in 2000, National Safety Council, Chicago, IL, 2000.
5. How to Reduce Workplace Accidents, European Agency for Safety and Health at Work, Brussels, Belgium, 2001.
6. Accident Facts, National Safety Council, Chicago, IL, 1990–1993.
7. Report on Injuries in America, National Safety Council, Chicago, IL, 2001.
8. Spellman, F.R. and Whiting, N.E., *Safety Engineering: Principles and Practice,* Government Institutes, Rockville, MD, 1999.
9. Blake, R.P., Ed., *Industrial Safety,* Prentice Hall, Englewood Cliffs, NJ, 1964.
10. Lancianese, F., The soaring costs of industrial accidents, *Occupational Hazards,* August 1983, pp. 30–35.
11. Hammer, W. and Price, D., *Occupational Safety Management and Engineering,* Prentice Hall, Upper Saddle River, NJ, 2001.
12. Operator safety, *Engineering,* May 1974, pp. 358–363.
13. Hunter, T.A., *Engineering Design for Safety,* McGraw-Hill, New York, 1992.
14. Dhillon, B.S., *Engineering Safety: Fundamentals, Techniques, and Applications,* World Scientific Publishing, River Edge, NJ, 2003.
15. Somers, A.R. and Somers, H.M., *Workmen's Compensation,* John Wiley & Sons, New York, 1945.
16. Eastman, C., *Work Accidents and the Law,* New York Charities Publication Committee, New York, 1910.
17. State Workers' Compensation Laws, The U.S. Department of Labour, Washington, D.C., January 1990.

13

Safety Analysis Methods

13.1 Introduction

Although the problem of safety has been around for a very long time, the development of safety analysis methods is relatively new. Some of these methods were specifically developed for application in safety area and the others for application in different areas. Some examples of these methods are failure modes and effect analysis (FMEA), Markov method, cause and effect diagram, and quality control charts. FMEA and the Markov method were developed for application in reliability areas and the cause and effect diagram and quality control charts for use in quality control work. Nonetheless, the main objective of all these safety analysis methods is to prevent the occurrence of accidents and hazards.

As the effectiveness of these methods can vary from one application to another, a careful consideration is necessary in selecting a method for a specific application. The safety analysis methods may be grouped under two main categories: mathematically based and nonmathematically based. For example, the Markov method and quality control charts belong to the mathematically based group and the cause and effect diagram and FMEA to the nonmathematically based category. This chapter presents important safety analysis methods taken from published literature [1–5].

13.2 Cause and Effect Diagram (CAED)

This method was developed by K. Ishikawa in Japan in the early 1950s for use in quality control work [6]. It can also be used to perform various types of safety analysis. The method is also known as *Fishbone diagram* and *Ishikawa diagram*. Visually, CAED may be divided into two parts: the right-hand side and the left-hand side. The right-hand side, i.e., the rectangle or the fish head, represents effect and left of it (i.e., the left-hand side) all possible causes that are connected to the central "fish" spine.

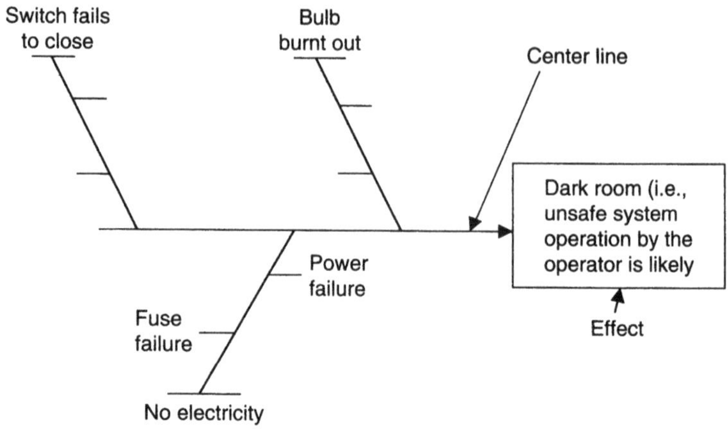

FIGURE 13.1
Cause and effect diagram for having dark room.

These causes can provide a useful checklist for the initial phase/stage safety analysis, and an appropriately developed diagram can serve as an important tool to identify safety-related problems. CAED is described in detail in Chapter 10. An application of CAED to a safety-related problem is demonstrated through the following example:

Example 13.1

An engineering system is located in a windowless room and is being operated by a human. For its safe operation by the human operator, the room must be lit all the time. The room has one light bulb and a switch. The switch can only fail to close and the room is supplied with regular electrical power only. Develop a cause and effect diagram for the effect: dark room (i.e., unsafe system operation by the operator is likely).

The cause and effect diagram is shown in Figure 13.1. In the diagram the causes for having dark room (effect) are no electricity, bulb burnt out, and switch fails to close. In turn, the subcauses of the cause (i.e., no electricity) are power failure and fuse failure.

13.3 Fault Tree Analysis (FTA)

This method was developed in the early 1960s at Bell Telephone Laboratories to perform reliability and safety analyses of the Minuteman Launch Control System. Today, FTA is used widely in industry to perform various types of reliability and safety analyses.

FTA begins by identifying an undesirable event, known as top event, associated with a system under consideration. Events which could cause the occurrence of the top event are generated and connected by logic operators such as OR and AND. The OR gate produces a true (say, unsafe) output if and only if one or more inputs are true (i.e., faulty or unsafe). The AND gate provides a true (say, unsafe) output if and only if all the inputs are true (i.e., faulty or unsafe).

The fault tree construction proceeds by generation of events (i.e., faulty or unsafe) successively until the events need not be developed any further. These events are known as primary (fault or unsafe) events. A fault tree itself is a logic structure that graphically relates the top event to the primary events. Moreover, in its construction involves successively asking the question "How could this event occur?" FTA is described in detail in Chapter 6.

The application of FTA to a safety problem is demonstrated through the following example:

Example 13.2

Assume that in Example 13.1 the room has four independent and identical light bulbs. The probability of performing an unsafe operation by the operator increases quite dramatically, if there is total darkness in the room. Develop a fault tree for the top event: dark room (i.e., unsafe system operation by the operator is likely), by using the fault tree symbols given in Chapter 6.

A fault tree for the example is shown in Figure 13.2.

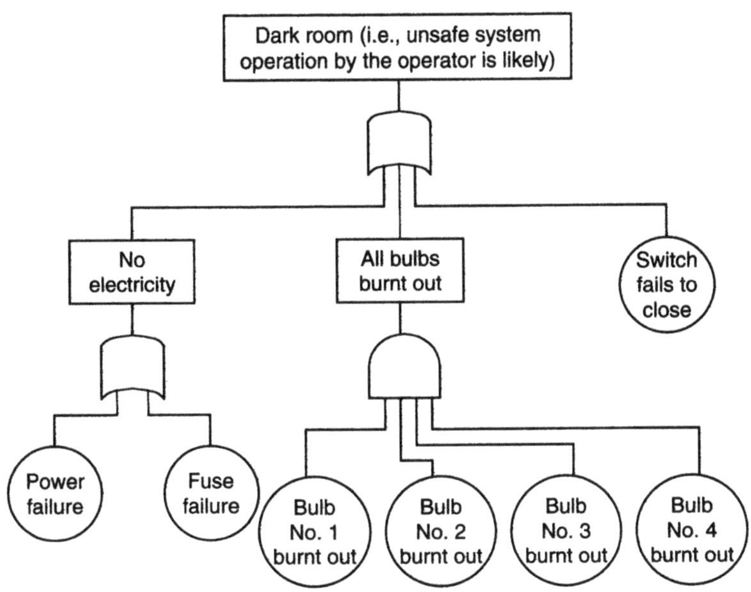

FIGURE 13.2
A fault tree for Example 13.2.

13.4 Control Charts

There are many types of control charts. They were originally developed by Walter A. Shewhart in 1924 for use in quality control work [7]. They can also be used to perform various types of safety analysis. A control chart is a graphical technique used to evaluate whether a given process is in a "state of statistical control" or out of control. More specifically, when a sample value falls outside the upper and lower control limits of a control chart, it means that the process is out of statistical control and requires an investigation.

In safety work, the process could be the frequency of accidents, severity of accidents, etc. There are many types of control charts that can be used in safety studies. Here, their application to safety-related problems is demonstrated through one type of control chart only (i.e., the C-chart). The C-chart is based on the Poisson distribution. The mean and standard deviation of the Poisson distribution are expressed by [8]

$$\mu = \frac{TA}{TP} \tag{13.1}$$

where

TA = total number of accidents
TP = total time period
μ = mean of the Poisson distribution

and

$$\sigma = (\mu)^{1/2} \tag{13.2}$$

where σ is the standard deviation of the Poisson distribution.

Thus, the upper and lower control limits of the C-chart are expressed by [9]

$$UCL_C = \mu + 3\sigma \tag{13.3}$$

and

$$LCL_C = \mu - 3\sigma \tag{13.4}$$

where UCL_C is the upper control limit of the C-chart and LCL_C is the lower control limit of the C-chart. The following example demonstrates the application of the C-chart to a safety problem.

Example 13.3

In an organization over a 12-month period, a total of 150 accidents occurred. Their monthly breakdowns are given in Table 13.1. Develop the C-chart and comment on it.

TABLE 13.1

Monthly Accident Occurrences

Month	No. of accidents
December (D)	18
November (N)	7
October (O)	16
September (S)	16
August (A)	9
July (J)	6
June (JE)	13
May (M)	17
April (AL)	11
March (MH)	12
February (F)	10
January (JY)	15

Substituting the given data into Equation (13.1) yields

$$\mu = \frac{150}{12} = 12.5 \text{ accidents/month}$$

Thus from Equation (13.2), we get

$$\sigma = (12.5)^{1/2} = 3.53$$

Substituting the above two calculated values into Equation (13.3) and Equation (13.4) yield

$$UCL_C = 12.5 + 3(3.53) = 23.10$$

and

$$LCL_C = 12.5 - 3(3.53) = 1.89$$

A C-chart for the Table 13.1 data and the above calculated values is shown in Figure 13.3. The chart shows that all the monthly accident occurrences are within the control limits; thus the process is in control.

13.5 Markov Method

This method is known after a Russian mathematician named Andrei A. Markov (1856–1922). It is widely used to perform reliability analysis of engineering systems when occurrence rates (i.e., failure and repair rates) are constant or

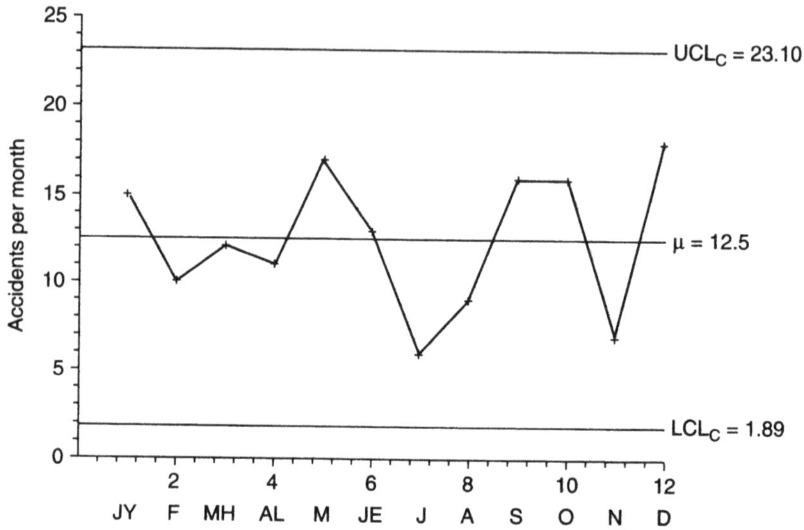

FIGURE 13.3
The C-chart for Example 13.3.

nontime dependent. The method can also be used to conduct various types of safety analysis. A detailed description of the method is given in Chapter 6.

The following example demonstrates the application of the Markov method to a safety problem.

Example 13.4

A system can either fail safely or unsafely. Both safe and unsafe failure rates are constant. The system state space diagram is shown in Figure 13.4. The numerals in box, diamond, and circle denote system states. Develop equations for system state probabilities and calculate the probability that during 500-h of operation the system will fail unsafely, if the system safe and unsafe failure rates are 0.009 failures/hour and 0.001 failures/hour, respectively. Assume that all system failures are statistically independent.

With the aid of the Markov method, we write down the following Equations for Figure. 13.4 diagram [5]:

$$\frac{dP_0(t)}{dt} + P_0(t)\lambda_1 + P_0(t)\lambda_2 = 0 \tag{13.5}$$

$$\frac{dP_1(t)}{dt} - P_0(t)\lambda_1 = 0 \tag{13.6}$$

$$\frac{dP_2(t)}{dt} - P_0(t)\lambda_2 = 0 \tag{13.7}$$

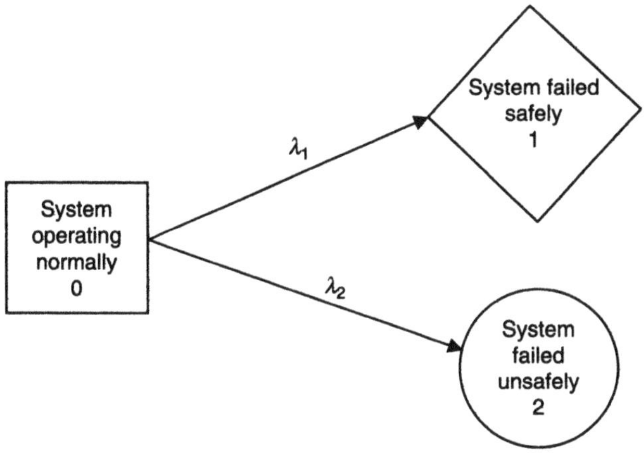

FIGURE 13.4
The system state space diagram.

At time $t = 0$, P_0 (0) = 1, P_1 (0) = 0, and P_2 (0) = 0. The symbols used in Equations (13.5–13.7) are defined below.

P_i (t) = probability that the system is in state i at time t, for $i = 0$ (operating normally), $i = 1$ (failed safely), and $i = 2$ (failed unsafely).
λ_1 = system constant safe failure rate.
λ_2 = system constant unsafe failure rate.

By solving Equations (13.5–13.7), we obtain

$$P_0(t) = e^{-(\lambda_1+\lambda_2)t} \tag{13.8}$$

$$P_1(t) = \frac{\lambda_1}{\lambda_1+\lambda_2}\left[1-e^{-(\lambda_1+\lambda_2)t}\right] \tag{13.9}$$

$$P_2(t) = \frac{\lambda_2}{\lambda_1+\lambda_2}[1-e^{-(\lambda_1+\lambda_2)t}] \tag{13.10}$$

Using the given data values in Equation (13.10) yields

$$P_2(500) = \frac{0.001}{(0.009+0.001)}[1-e^{-(0.009+0.001)(500)}]$$
$$= 0.0993$$

Equations (13.8–13.10) are for system state probabilities and the probability that during 500-h of operation the system will fail unsafely is 0.0993.

13.6 Failure Modes and Effect Analysis (FMEA)

This is a widely used technique to perform reliability analysis of engineering systems. It analyzes each potential failure mode in a given system to determine the effects of such failure modes on the entire system. FMEA can also be used to conduct various types of safety analysis.

The method is essentially composed of seven main steps [9]: (i) establish system definition (i.e., define the system and its associated requirements); (ii) establish ground rules for performing FMEA; (iii) develop system hardware description; (iv) develop description of system functional blocks; (v) identify failure modes and their effects for each and every item in the system; (vi) compile critical item list; and (vii) document the analysis. FMEA is described in detail in Chapter 6.

13.7 Hazards and Operability Analysis (HAZOP)

This method was developed for use in chemical industry [10–11]. It is extremely useful to identify safety-related problems prior to availability of full data concerning an item. Three fundamental objectives of the HAZOP are—produce a complete description of process/facility, review each process/facility element to determine how deviations from the design intentions can happen, and decide whether the deviations can lead to operating problems/hazards [1–5].

A HAZOP study is performed by following seven steps shown in Figure 13.5 [1, 12].

HAZOP has basically the same weaknesses as FMEA. For example, they both predict problems related to process/system failures, but do not factor human error into the equation. This is the key weakness because human error is frequently a factor in accidents.

13.8 Technic of Operations Review (TOR)

This is another safety analysis method and it was developed by D.A. Weaver of the American Society of Safety Engineers (ASSE) in the early 1970s [1]. It may simply be described as a hands on analytical methodology or approach developed for determining the root system causes of an operation failure or malfunction. The method uses a worksheet containing simple and straightforward terms requiring yes/no decisions. The basis for the activation of

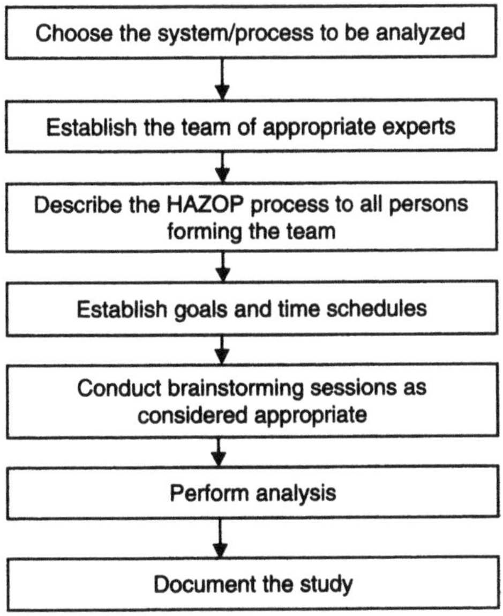

FIGURE 13.5
Steps for performing a HAZOP study.

TOR is an incident occurring at a certain time and place involving certain people. The following eight steps are associated with the TOR [1, 13]:

- Form the TOR team.
- Hold a roundtable session with all team members. The main objective of this session is to communicate common knowledge to all parties involved.
- Identify a single important factor that has played a pivotal role in the occurrence of incident/accident. This factor is the result of team consensus and serves as a starting point to all further investigations.
- Use the team consensus in responding to a sequence of yes/no options.
- Evaluate all identified factors and ensure the existence of consensus among the members of the team.
- Prioritize all the contributing factors under consideration.
- Develop appropriate preventive/corrective strategies in regard to each and every contributing factor under consideration.
- Implement strategies.

All in all, the strength of TOR stems from the involvement of line personnel in the analysis and its weakness from the fact that it is specifically designed as an after-the-fact-process.

13.9 Interface Safety Analysis (ISA)

This method is concerned with determining the incompatibilities between subsystems and assemblies of a given item/product that may result in accidents. More specifically, the method establishes that distinct items/parts/units can be integrated into a viable system and an individual unit's or part's normal operation will not impair the performance or cause damage to another part/unit or the overall system. Although, ISA considers various relationships but they can be classified under three groups shown in Figure 13.6 [14].

Functional relationships are associated with multiple items or units. For example, in a situation where outputs of an item/unit constitute the inputs to a downstream item/unit, any error in outputs and inputs may cause damage to the downstream item/unit and in turn a safety hazard.

Flow relationships are associated with two or more units. Moreover, the flow between two units may involve items such as air, water, electrical energy, steam, fuel, or lubricating oil. Usually, the common problems associated with many products are the effective flow of energy and fluids from one item to another through confined passages, thus resulting in direct/indirect safety-related problems. All in all, with respect to fluids, factors such as loss of pressure, contamination, toxicity, corrosiveness, lubricity, and flammability must be considered with care from the safety aspect.

Physical relationships are associated with the physical aspects of units or items. For example, some units or items may function well individually, but they may fail to fit together because of dimensional differences or other problems. In turn, this may lead to various types of safety problems.

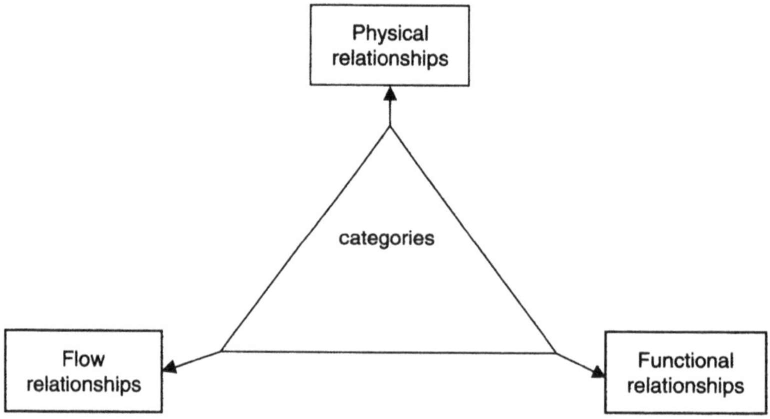

FIGURE 13.6
Categories of relationships considered by ISA.

13.10 Job Safety Analysis (JSA)

This method is concerned with finding and rectifying potential hazards that are intrinsic to or inherent in a given workplace. Usually, individuals such as worker, supervisor, and safety professional participate in JSA. The following steps are associated with the performance of JSA [15]:

- Select a job for analysis.
- Break down the job into various tasks/steps.
- Identify all possible hazards and propose suitable measures for controlling them to appropriate levels.
- Apply the proposed measures.
- Evaluate the end results.

Past experiences indicate that the success of JSA depends on the degree of rigor exercised by the JSA team during the analysis process.

13.11 Safety Indexes

These indexes are concerned with measuring the safety performance of an organization. Over the years many such indexes have been developed [16]. Two safety indexes proposed by the American National Standards Institute are as follows [17]:

13.11.1 Disabling Injury Severity Rate (DISR)

This is defined as follows:

$$\text{DISR} = \frac{D_C(1{,}000{,}000)}{\text{EET}} \tag{13.11}$$

where *EET* is the employee exposure time expressed in hours and D_C is the total number of days charged.

The index is based on four factors occurring during the period covered by the rate (i.e., total scheduled charges (days) for all deaths, permanent total, and permanent partial disabilities, and the total number of days of disability from all temporary injuries). Some of the advantages of the index are as follows:

- Useful to make a meaningful comparison between different organizations.

- A useful tool to answer the question: "how serious are injuries in our organization?"
- A useful tool to take into consideration differences in quantity of exposure over time.

13.11.2 Disabling Injury Frequency Rate (DIFR)

This is defined by

$$DIFR = \frac{TNDI(1{,}000{,}000)}{EET} \tag{13.12}$$

where TNDI is the total number of disabling injuries.

The index is based on a total of four events that occur during the time period covered by the rate (i.e., deaths, permanent disabilities, permanent partial disabilities, and temporary disabilities).

One important benefit of this index is that it considers differences in quantity of exposures due to varying employee/worker work hours, either within the framework of the organization during successive periods or among organizations classified under the similar industry group [16].

13.12 Problems

1. Describe the cause and effect diagram.
2. What are the other names used for the cause and effect diagram?
3. Make a comparison between fault tree analysis and failure modes and effect analysis.
4. Write an essay on the use of control charts in safety work.
5. Prove that in Example 13.4 the probability of the system failing unsafely is given by Equation (13.10).
6. Discuss the seven main steps used in performing failure modes and effect analysis.
7. Describe hazards and operability analysis (HAZOP).
8. List the steps used in performing technic of operations review (TOR).
9. Discuss the following two items:
   - Interface safety analysis
   - Job safety analysis
10. What are the principal advantages of the disabling injury severity rate (DISR)?

References

1. Goetsch, D.L., *Occupational Safety and Health*, Prentice Hall, Englewood Cliffs, NJ, 1996.
2. Roland, H.E. and Moriarty, B., *System Safety Engineering and Management*, John Wiley & Sons, New York, 1983.
3. Gloss, D.S. and Wardle, M.G., *Introduction to Safety Engineering*, John Wiley & Sons, New York, 1984.
4. Raouf, A. and Dhillon, B.S., *Safety Assessment: A Quantitative Approach*, Lewis Publishers, Boca Raton, FL, 1994.
5. Dhillon, B.S., *Engineering Safety: Fundamentals, Techniques, and Applications*, World Scientific Publishing, River Edge, NJ, 2003.
6. Mears, P., *Quality Improvement Tools and Techniques*, McGraw-Hill, New York, 1995.
7. *Statistical Quality Control Handbook*, AT & T Technologies, Indianapolis, IN, 1956.
8. Ireson, W.G., Ed., *Reliability Handbook*, McGraw-Hill, New York, 1966.
9. Jordan, W.E., Failure modes, effects and criticality analysis, *Proceedings of the Annual Reliability and Maintainability Symposium*, 1972, pp. 30–37.
10. *Guidelines for Hazard Evaluation Procedures*, American Institute of Chemical Engineers, New York, 1985.
11. Dhillon, B.S. and Rayapati, S.N., Chemical systems reliability: a survey, *IEEE Transactions on Reliability*, Vol. 37, 1988, pp. 199–208.
12. Risk Analysis Requirements and Guidelines, Report No. CAN/CSA-Q634-91, prepared by the Canadian Standards Association, 1991. Available form Canadian Standards Association, 178 Rexdale Boulevard, Rexdale, Ontario, Canada.
13. Hallock, R.G., Technic of operations review analysis: determine cause of accident/incident, *Safety and Health*, Vol. 60, August 1991, pp. 38–39, 46.
14. Hammer, W., *Product Safety Management and Engineering*, Prentice Hall, Englewood Cliffs, NJ, 1980.
15. Hammer, W. and Price, D., *Occupational Safety Management and Engineering*, Prentice Hall, Upper Saddle River, NJ, 2001.
16. Tarrants, W.E., *The Measurement of Safety Performance*, Garland STPM Press, New York, 1980.
17. Z-16.1, Method of Recording and Measuring Work Injury Experience, American National Standards Institute, New York, 1985.

14

Safety Management and Costing

14.1 Introduction

Safety management is an important element of a safety program in an organization. The beginning of the safety management appears to be the period during the 1950s and 1960s [1]. An important milestone in the history of safety management occurred in 1970 with the passage of the Occupational Safety and Health Act (OSHA) by the U.S. Congress. Since 1970, many new developments related to safety management have occurred. Safety management may simply be defined as the accomplishment of safety through the efforts of others [1–4]. Nonetheless, the fundamental objective of safety management is to eliminate human suffering and anguish as well as to achieve economy of operations.

Just like in any other area of engineering, cost is also an important factor in safety. In fact, it was basically due to the cost of safeguarding systems, the factor of safety was generally overlooked during the early years of the Industrial Revolution. For example, once a railroad executive remarked: "It would cost less to bury a person killed in an accident than to put air brakes on a car" [5]. Today, one of the major factors for the increasing attention on safety is the cost of accidents. This cost includes items such as medical expenses, wage losses, property damage amount, lawsuit expenses, and productivity losses.

This chapter presents various important aspects of safety management and costing.

14.2 Safety Management Principles and Developing a Safety Program Plan

There are many safety management principles. Some of the important ones are as follows [6–7]:

- The function of safety is to find and define accident causing operational errors.

- The safety system should be tailored to fit the organization/company culture.
- Management should manage safety just like managing any other function in the organization (i.e., by setting attainable goals, planning, organizing, and controlling).
- In developing an effective safety system carefully consider three major subsystems, i.e., the managerial, the physical, and the behavioral.
- Management has the responsibility to make changes to the environment that leads to the unsafe behavior because unsafe behavior is the result of normal people reacting to their surrounding environment.
- A key for having an effective safety system is flexibility, worker participation, visible top management support, etc.
- Causes that lead to unsafe behavior can be identified, classified, and controlled.
- Symptoms that indicate something is wrong in the safety management system are an unsafe condition, an unsafe act, and an accident.
- Circumstances that can be predicted to result in serious injuries are non-productive activities, nonroutine activities, certain construction situations, and high energy sources.
- The key to successful line safety performance is management procedures that factor in accountability in an effective manner.

The development of a safety plan is essential for organizations contemplating on introducing a safety program. A number of useful steps for developing such a plan are as follows [2]:

- **Develop and announce the safety policy.** This step is concerned with preparing and announcing the policy for controlling hazards within the organization and designating authority and accountability for its implementation.
- **Appoint safety chief.** This step is concerned with appointing an individual with proper qualifications and experience to look after all safety-related matters within the organization.
- **Analyze operational injury records.** This step is concerned with performing analysis of operational records of injuries, work-related illnesses, and property damage.
- **Evaluate operational hazards' scope and seriousness.** This step is concerned with items such as evaluating the quality of the existing physical safeguards, determining the nature and severity of inherent operating hazards, determining the corrective measures required, and

developing time estimates and budgets for performing the corrective measures.

- **Select and schedule appropriate communication methods.** This step is concerned with selecting and scheduling appropriate communication approaches for various purposes including providing safety training to workers and informing higher management about the organization's safety progress and associated requirements.
- **Establish a schedule for periodic reviews.** This step is concerned with developing a schedule for periodic reviews of the program and facilities with respect to safety.
- **Establish objectives.** This step is concerned with developing appropriate short-range and long-range objectives for the safety program.

14.3 Safety Department Functions

A safety department performs various types of functions. These functions can vary from one organization to another. Nonetheless, some of the typical functions of a safety department are shown in Figure 14.1 [2, 8].

14.4 Functions and Qualifications of Safety Professionals

Two key professionals that play an instrumental role in an organization's safety program are the safety manager and safety engineer. Functions and qualifications of each of these two individuals are presented below separately.

14.4.1 Safety Manager

This individual must possess certain qualifications in order to carry out his/her task in an effective manner. Some of these qualifications are appropriate academic qualifications, effective knowledge in safety and safety related experience, good knowledge of business administration, effective knowledge in engineering principles, open mind to try new ideas and methods, good drive, enthusiasm, and perseverance; ability to get others to do their tasks in an effective manner, and good supervisory skills [2, 6, 9].

A safety manager performs various functions including supervising department employees, managing and formulating the safety program, participating in procurement specification reviews, acquiring the latest hazard control

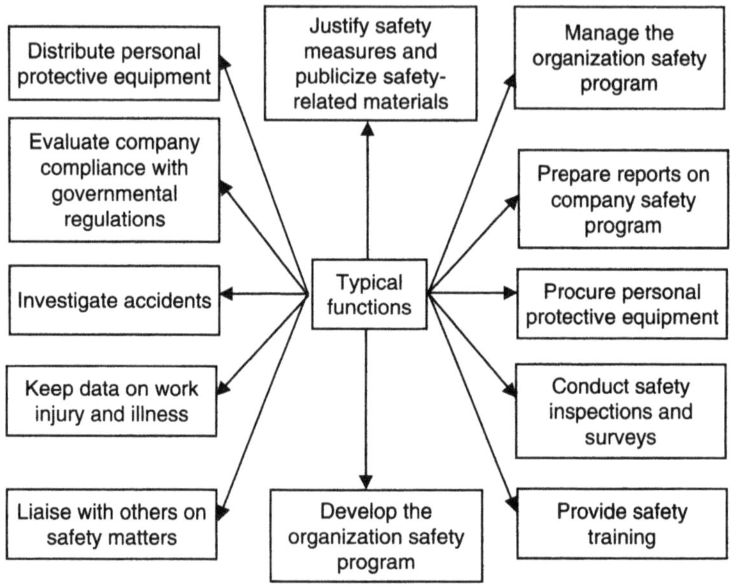

FIGURE 14.1
Typical functions of a safety department.

information, participating in the design of new facilities/equipment layout/process layout, reporting to higher management periodically with respect to the state of the company's safety effort, representing management to public, government agencies, insurance companies, etc. with respect to safety, directing the collection and recording of pertinent information on matters such as work injuries and accidents, advertising safety issues at all levels of management, and directing the inspection of the facilities for compliance with the regulations of the outside bodies [2, 6, 9].

14.4.2 Safety Engineer

A safety engineer must possess certain academic and other qualifications in order to perform his/her job effectively. In 1992, the American Society of Safety Engineers (ASSE) conducted a survey of its 27,000 members with regard to their academic qualifications. The results of the survey are shown in Figure 14.2 [10].

The other qualifications include readiness to try new ideas and approaches, enthusiasm, drive, perseverance, easy to get along with safety related experience, and a good knowledge of safety matters.

Some of the functions performed by a safety engineer are conduct safety inspections, collaborate with safety committees, perform safety studies, conduct accident investigations, coordinate with management on matters

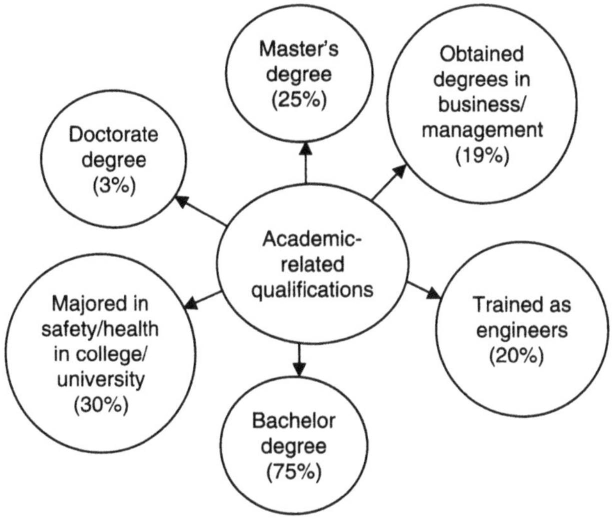

FIGURE 14.2
Academic-related qualifications of the American Society of Safety Engineers Members.

related to safety, provide safety training, process workers' compensation claims, and ensure that the appropriate corrective measures are taken to avoid accident re-occurrences [8].

14.5 Safety Committees and Motivating Employees to Work Safely

Safety committees play an important role in promoting safety within an organization. They provide a systematic structure for funnelling health and safety issues by workers and management. Prior to starting a safety committee careful attention must be paid to factors such as scope/responsibility of the committee, authority of the committee, and procedures. The procedures are concerned with items such as listed below [2, 10].

- Meeting frequency
- Records to be retained
- Attendance requirements
- Time and place of meetings
- The authority to whom committee reports are to be submitted

Past experiences indicate that success or failure of a safety committee largely depends on its composition. It means that for the success of committees, the committee members must be from a broad cross-section representing all units within the organization. This will provide all employees a representative voice on the committee.

Nonetheless, the health and safety professional member of the committee serves as a facilitator, advisor, and catalyst. The members of the committee appoint a chairperson and a secretary for recording the minutes. Past experiences indicate that usually best results are obtained, when neither the health and safety professional nor the executive manager serves as a committee chairperson. In addition, all the safety committee members are trained in areas such as group problem-solving methods and cause-and-effect analysis [11].

Some of the important duties of a safety committee are providing assistance in safety training, providing assistance to safety professionals, promoting accident prevention, conducting inspection tours, making safety recommendations, providing assistance in safety deficiency investigations, and providing assistance in selecting safety-related posters [5].

Supervisors play a key role in motivating employees to work safely. They can motivate employees directly or indirectly through actions such as listed below [7].

- Focusing attention on the importance of safe performance.
- Reminding workers about the methods of safe performance.
- Spending more time recognizing and rewarding safe performance than spending time in disciplining employees for unsafe performance.
- Strengthening and enhancing the importance of management standards for safe performance.

14.6 A Manufacturer's Losses or Cost due to an Accident Involving its Product

A manufacturer's losses or cost due to an accident involving its product occur in many different ways. Some of these are as follows [12]:

- Payments for injury/death claim settlements.
- Cost of legal aid for defense against claims.
- Cost of accident investigation.
- Cost of corrective measures required to prevent accident reoccurrences.
- Lost time of manufacturer's personnel involved with the accident.
- Increase in insurance cost.

- Cost of slowdown in operations while accident causes are being investigated.
- Payments for property damage claims that are not covered by insurance.
- Loss of revenue because of degradation in public confidence.

14.7 Safety Cost Estimation Methods and Models

There are many methods and models used to estimate various types of safety costs. This section presents some of these methods and models.

14.7.1 The Simonds Method

This method was developed by Professor R.H. Simonds of Michigan State College by reasoning that the cost of an accident is composed of two main elements, i.e., insured cost and uninsured cost [13]. Furthermore, he argued that the insured cost can easily be estimated by simply examining some accounting records or data but the estimation of the uninsured cost is much more challenging or a trying task. Thus, for estimating the total uninsured cost of accidents, Simonds proposed the following relationship [3, 13–14]:

$$\text{TUCA} = \alpha_A AUC_A + \alpha_B AUC_B + \alpha_C AUC_C + \alpha_D AUC_D \tag{14.1}$$

where

TUCA = total uninsured cost of accidents

AUC_A = average uninsured cost associated with Class A accidents

α_A = total number of lost work-day cases due to Class A accidents resulting in permanent partial disabilities and temporary total disabilities

AUC_B = average uninsured cost associated with Class B accidents

α_B = total number of physician's cases associated with Class B accidents. More specifically, the Occupational Safety and Health Act (OSHA) nonlost work day cases that require treatment by a doctor.

AUC_C = average uninsured cost associated with Class C accidents

α_C = total number of first-aid cases associated with Class C accidents. More specifically, those accidents in which first-aid was provided locally and resulting in a loss of less than one eighth of working time

AUC_D = average uninsured cost associated with Class D accidents

α_D = total number of noninjury cases associated with Class D accidents. More specifically, those accidents causing only minor injuries that do not need the attention of a medical professional

14.7.2 The Heinrich Method

This method is named after H.W. Heinrich who categorically pointed out that for each and every insured cost dollar paid for accidents, there were around four dollars of uninsured cost borne by the organization or company [15]. His conclusions were based on factors such as listed below [16].

- Examination of a total of 5000 case files from organizations insured with a private company.
- Performance of research studies in those organizations.
- Interviews with the administrative and production service staff of these enterprises.

Heinrich defined "total cost of occupational injuries" as the sum of two elements: direct cost and indirect cost. The direct cost is the total benefits paid by the insurance company whereas the expenditure assumed directly by the enterprise is the indirect cost. More specifically, the direct cost is composed of elements such as cost of injured worker's lost time, cost of lost orders, cost of machine/material damage, cost of management's lost time, lost time cost of workers who stop their work and get involved in the action, cost of weakened morale, lost time cost of first aid and hospital workers not paid by insurance, cost associated with profit and worker productivity loss, and cost of overheads for injured worker while in nonproduction mode.

14.7.3 Total Safety Cost Estimation Model

In this model, the total safety cost is expressed by [5]:

$$SC_t = CI + CILDA + CWI + CAPM + CIM + CSLI + CRR + CMSI \tag{14.2}$$

where

CI = cost of insurance
CILDA = cost of immediate losses due to accidents
CWI = cost of welfare issues
CAPM = cost of accident prevention measures
CIM = cost of immeasurables
CSLI = cost of safety-related legal issues
CRR = cost of rehabilitation and restoration
CMSI = cost of miscellaneous safety issues

14.8 Safety Cost Indexes

These indexes are used to measure the overall safety cost performance of organizations. The main objective of using these indexes is to indicate trends, using the past safety cost performance as a point of reference, and to encourage all involved individuals to improve over the past performance.

This section presents three most useful safety cost-related indexes taken from published literature [14, 16, 17].

14.8.1 Index I

This index is concerned with determining the average cost of occupational injuries per profit dollar in an organization. The index is defined by

$$AC_{pd} = \frac{TC_{oi}}{TP} \tag{14.3}$$

where

AC_{pd} = average cost of occupational injuries per profit dollar
TP = total profit expressed in dollars
TC_{oi} = total cost of occupational injuries expressed in dollars

14.8.2 Index II

This index is concerned with determining the average cost of occupational injuries per unit turnover. The index is defined by

$$AC_{ut} = \frac{TC_{oi}}{TNUT} \tag{14.4}$$

where AC_{ut} is the average cost of occupational injuries per unit turnover and TNUT is the total number of units turnover (i.e., unit quantity, tons, etc.).

14.8.3 Index III

This index is concerned with determining the average cost per occupational injury in an organization. The index is defined by

$$AC_{oi} = \frac{TC_{oi}}{TNOI} \tag{14.5}$$

where AC_{oi} is the average cost per occupational injury and TNOI is the total number of occupational injuries.

14.9 Problems

1. Discuss principles of safety management.
2. Define the term "Safety management."
3. Discuss the steps involved in developing a safety program plan.
4. List at least ten important functions of a safety department.
5. What are the important functions of a safety manager?
6. What are the important functions of a safety engineer?
7. What are the important duties of a safety committee?
8. Discuss the following safety cost estimation methods:
    - The Simonds method
    - The Heinrich method
9. Define at least two indexes used in measuring the overall performance of organizations with respect to injury cost.
10. What are a manufacturer's losses due to an accident involving its product?

References

1. Petersen, D., *Safety Management: A Human Approach*, Aloray Publishers, Englewood Cliffs, NJ, 1975.
2. Grimaldi, J.V. and Simonds, R.H., *Safety Management*, Richard D. Irwin, Chicago, 1989.
3. Goetsch, D.L., *Occupational Safety and Health*, Prentice Hall, Englewood Cliffs, NJ, 1996.
4. Roland, H.E. and Moriarty, B., *System Safety Engineering and Management*, John Wiley & Sons, New York, 1983.
5. Hammer, W. and Price, D., *Occupational Safety Management and Engineering*, Prentice Hall, Upper Saddle River, NJ, 2001.
6. Dhillon, B.S., *Engineering Safety: Fundamentals, Techniques, and Applications*, World Scientific Publishing, River Edge, NJ, 2003.
7. Petersen, D., *Safety Management*, American Society of Safety Engineers, Des Plaines, IL, 1998.
8. Gloss, D.S. and Wardle, M.G., *Introduction to Safety Engineering*, John Wiley & Sons, New York, 1984.
9. Schenkelbach, L., *The Safety Management Primer*, Dow Jones-Irwin, Homewood, IL, 1975.
10. Blake, R.P., Ed, *Industrial Safety*, Prentice Hall, Englewood Cliffs, NJ, 1964.
11. Peters, T., *Thriving on Chaos: Handbook for a Management Revolution*, Harper and Row, New York, 1987.
12. Hammer, W., *Product Safety Management and Engineering*, Prentice Hall, Englewood Cliffs, NJ, 1980.

13. Simonds, R.M., Estimating accident cost in industrial plants, Safety Practices Pamphlet No. III, National Safety Council, Chicago, 1950.
14. Raouf, R., Dhillon, B.S., *Safety Assessment: A Quantitative Approach*, Lewis Publishers, Boca Raton, FL, 1994.
15. Heinrich, H.W., *Industrial Accident Prevention*, McGrawHill, New York, 1931.
16. Andreoni, D., The cost of occupational accidents and diseases, International Labour Office, Geneva, Switzerland, 1986.
17. Blake, R.P., Ed, *Industrial Safety*, Prentice Hall, Englewood Cliffs, NJ, 1964.

15

Robot, Software, and Medical Device Safety

15.1 Introduction

Each year a vast sum of money is being spent worldwide to design, develop, and operate robots, software, and medical devices. Their safety has become an important issue.

The word "robot" is derived from the Czech language, in which it means "worker." Today, there are over one million robots in use worldwide. The awareness of robot safety appears to have its beginning in the 1970s [1]. In the 1980s, three important documents concerning robot safety appeared: a technical guidance document [2], a book (edited) [3], and a standard [4]. In the 1990s, a book entitled "Robot Reliability and Safety" discussed the subject of robot safety in considerable depth [1]. A comprehensive list of publications on robot safety is given in Ref. [5].

Today, computers are used widely in various types of applications and they are composed of both hardware and software elements. In fact, nowadays much more money is spent in developing computer software than in developing hardware, in comparison to the early computers. For example, in 1955 software accounted for about 20% of the total computer cost and in the mid-1980s, the percentage skyrocketed to approximately 90% [6]. Needless to say, safety has become a critical issue in software because proper functioning of software is so important that a simple malfunction can result in a large-scale loss of lives. For example, commuter trams in Paris, France serve approximately 800,000 passengers daily and very much depends on software signalling [7].

Safety has also become an important issue in medical systems/devices as around 100,000 patients die annually due to avoidable mistakes in the U.S. alone [8]. Three legislations that have played an important role in promoting medical system/device safety in the U.S. were the Occupational Safety and Health Act (OSHA), 1970; the Medical Device Amendments, 1976; and the Safe Medical Device Act (SMDA), 1990 [9].

This chapter presents various important aspects of robot, software, and medical device safety.

15.2 Robot Safety

Robot safety may simply be described as preventing the robot from damaging its environment, particularly the human element of that environment or simply preventing damage to the robot itself [10]. Five important aspects of robot safety are presented below.

15.2.1 Facts and Figures

Some of the directly or indirectly robot safety-related facts and figures are as follows [11]:

- In 1978, the first robot-induced fatal accident occurred in Japan [12].
- A study of 32 robot-related accidents in the U.S., Japan, Sweden, and Germany reported that line workers were at the greatest risk of injury followed by maintenance personnel [13].
- In 1984, the first fatal robot accident occurred in the U.S. [14].
- During 1978–1984, a total of five fatal robot accidents occurred (i.e., four in Japan and one in the U.S.) [15].
- A study reported that approximately 12–17% of the accidents in the industrial sector using advanced manufacturing technology were associated with automated production equipment [16, 17].
- During 1978–1987, a total of ten fatal robot accidents occurred in Japan alone [12].

15.2.2 Robot Safety Problems

Over the years professionals working in the area of robotics have identified many safety problems unique to robots. Some of these problems are as follows [18]:

- The presence of a robot usually receive a great deal of attention from humans. These humans are often quite ignorant of the potential associated hazards.
- A robot may lead to a high risk of fire or explosion if it is installed in unsuitable environments.
- A robot creates potentially hazardous situations because it often manipulates objects of different weights and sizes.
- A robot may go out of its programmed zones, strike something or throw objects in the event of a control, hydraulic, or mechanical failure.
- Mechanical design-associated problems of robots may lead to hazards such as grabbing, pinning, and pinching.

- Robots are prone to occurrence of safety-related electrical design problems such as potential electric shock, poorly designed power sources, and fire hazards.
- Robot maintenance procedures can lead to various hazardous conditions.

15.2.3 Types of Robot Accidents

Robot accidents may be grouped under the following four distinct categories [19]:

- **Collision/impact accidents.** These are concerned with situations where unpredicted movements, part failures, or unpredicted program changes related to the arm of robot/peripheral equipment result in contact accidents.
- **Mechanical part accidents.** These include conditions where the breakdown of the robot's peripheral equipment, drive components, power source, or its tooling or end-effectors occurs. Two typical examples of mechanical failures are the failure of gripper mechanism, and the failure of end-effectors or power tools.
- **Trapping/crushing accidents.** These are concerned with conditions where an individual's limb or other body part is trapped between peripheral equipment and the robot arm or the individual is physically driven into and crushed by other peripheral equipment.
- **Miscellaneous accidents.** These are all those accidents that cannot be classified under the above three categories. Some examples of these accidents are environmental accidents from arc flash, dust, metal spatter, electromagnetic, or radio-frequency interference and tripping hazards from equipment and power cables on the floor.

15.2.4 Robot Hazard Causes

Over the years professionals working in the area of robotics have studied robot hazards to identify their causes. They have identified the following major causes for the occurrence of various types of robot hazards [19, 20]:

- **Unauthorized access.** This is concerned with the unauthorized entry into the safeguarded robot zone by an individual who may be unfamiliar about the safeguards in place or the activation status.
- **Control errors.** These errors can occur due to factors such as faults in the hydraulic, pneumatic, or electrical subcontrols associated with the robot system.
- **Human errors.** These are an important cause of robot hazards. Some examples of the robot-associated human errors are incorrect activation of the "teach pendant" or control panel, connecting live

input-output sensors to the microprocessor or a peripheral, and placing oneself in a hazardous position while programming the robot or performing maintenance on it because of over familiarity.

- **Improper installation.** It can cause various types of robot hazards. For example, the requirement design, equipment layout, utilities, and facilities of a robot system, if executed incorrectly, can lead to various inherent hazards.
- **Power systems.** These can also cause various types of robot hazards because pneumatic, electrical, or hydraulic power sources with malfunctioning control or transmission components in the robot power system can cause a disruption in electrical signals to the control/power-supply lines.
- **Mechanical failures.** These can cause robot hazards because operating programs may not account for cumulative mechanical part failure, thus contributing to the occurrence of erroneous or unexpected robot operation.
- **Environmental sources.** These can also be a cause of various types of robot hazards. For example, radio frequency/electromagnetic interference (transient signals) can exert an undesirable influence to a certain degree on robot operation, thus increasing the probability of injury to humans working in the area.

15.2.5 Robot Safeguard Methods

There are many robot safeguard methods. Some of the commonly used such methods are shown in Figure 15.1 [15, 21]. The flashing lights safeguard method is concerned with installing flashing lights at the perimeter of the

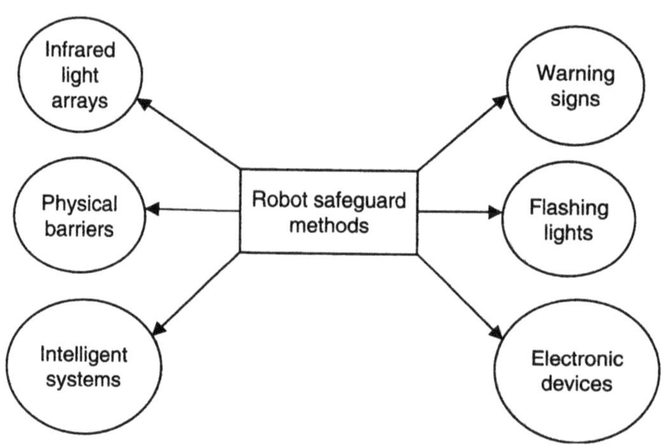

FIGURE 15.1
Common robot safeguard methods.

robot-working area or the robot itself. Their purpose is to alert humans that robot programmed motion is happening or could happen at any moment. The electronic devices approach is concerned with making use of active sensors for intrusion detection. Usually, this method is used in situations where the need of unobstructed floor space is important.

The warning signs approach is usually used in circumstances where robots, by virtue of their size, speed, and inability to impart excessive force, cannot injure people. The infrared light arrays are an effective approach used for protecting humans from potential dangers in the operating zone of a robot. The commonly used linear arrays of infrared sources are known as light curtains. Although, this approach has proven to be quite effective but time to time false triggering may occur due to factors such as smoke, heavy dust, or flashing lights because of misalignment of system elements.

The intelligent systems method uses intelligent control systems for safeguarding. More specifically, these systems make use of avenues such as remote, sensing, software, and hardware in making decisions. Needless to say, an intelligent collision-avoidance system can only be achieved by restricting the operating environment of robots and using special software and sensors. The physical barriers approach uses items such as chain-link fences, plastic safety chains, and safety rails to stop people from entering the restricted robot work zone.

15.3 Software Safety

Software safety may simply be described as the freedom from software-related hazards. This section presents five important aspects of software safety.

15.3.1 Facts and Figures

Some of the facts and figures directly or indirectly related to software safety are as follows [11]:

- A software error in a computer-controlled therapeutic radiation machine called Therac 25 resulted in deaths of two patients and severe injuries to another patient [22–24].
- A SAAB JAS39 Gripen fighter plane crashed due to an instrument failure caused by a safety-related software issue [25].
- A software error in a French meteorological satellite caused the destruction of 72 of the 141 weather balloons [26].
- Over 70% of the companies involved in software development develop their software products by using ad hoc and unpredictable approaches [27].
- A software error caused a radioactive heavy water spill at a Canadian nuclear power generating station [25].

15.3.2 Software Safety vs. Reliability

Often software safety and reliability are equated; thus leading to a considerable degree of confusion. In order to avoid such confusion, their clear understanding is important. In a broader perspective, it may simply be stated that the reliability requirements are concerned with making a system failure free and the safety mishap free [28, 29]. More specifically, reliability is concerned with all possible software errors and the safety with only those errors that may result in system hazards.

Nonetheless, it may be said that not all software functioning as per specification is safe nor all soft-errors cause safety problems [11]. For example, in the past severe mishaps have occurred while something (i.e., software) was operating exactly as per the specified requirement (i.e., without failure) [30, 31].

15.3.3 Software Hazard Causing Ways

Software can cause or contribute to a hazards in many different ways. Some of the important ones are as follows [28, 32]:

- By performing a function out-of-sequence
- By performing a function that is not required
- By providing a wrong solution to a given problem
- By failing to perform a required function
- By responding poorly to a contingency
- By failing to recognize a hazardous situation requiring a corrective action
- By poorly timing a response for an adverse situation

15.3.4 Basic Software System Safety Tasks

There are many software system safety tasks. Some of the basic ones include identifying the safety-critical elements and variables for use by code developers, tracing safety requirements and constraints right up to the code, performing any special safety analyses (e.g., software fault tree analysis or computer-human interface analysis), reviewing the test results concerning safety issues and tracing the highlighted safety-associated software problems back to the system level, tracing the identified system hazards to the hardware-software interface, developing safety-associated software test plans, procedures, case requirements, and descriptions, identifying those elements of the software that control safety-critical operations, and showing the software system safety constraint consistency with respect to the software requirements specification [33].

15.3.5 Software Hazard Analysis Methods

There are a large number of methods and techniques that can be used to perform various types of software hazard analysis [34–35]. These include code walk-throughs, failure modes and effect analysis (FMEA), proof of correctness, software sneak circuit analysis, design walk-throughs, Petri net analysis, structural analysis, cross-reference-listing analysis, event-tree analysis, cause-consequence diagrams, hazard and operability studies, nuclear safety cross-check analysis, and software fault tree analysis.

Three of these methods are described below [36].

15.3.5.1 Proof of Correctness

This method decomposes a given program into a number of logical segments and defines each segment input/output assertions. Subsequently, software professionals verify from the perspective that each and every input assertion and its associated output assertion are true and that if all of the input assertions are true, then all output assertions are also true.

Furthermore, the proof of correctness approach makes use of mathematical theorem proving concepts to verify that a given program is consistent with its associated specifications. The method is described in Refs. [34–37].

15.3.5.2 Code Walk-Throughs

This is an effective approach used for improving software safety and quality and basically is a team effort among various professionals: software engineers, system safety specialists, software programmers, program managers, etc.

Nonetheless, code walk-throughs are in-depth review of a given software through inspection and discussion of its functionality. More specifically, all logic branches and the function of each statement are discussed in a significant depth. Thus, the process provides a checks and balances system of the software produced. Additional information on the method is available in Ref. [37].

15.3.5.3 Software Sneak Circuit Analysis

This method identifies software logic that causes undesired outputs. More specifically, program source code is converted to topological network trees, and patterns such as entry dome, return dome, iteration loop, trap, single line, and parallel line are used to model the code. Each and every software mode is modeled using the basic patterns linked in a network tree flowing from top to bottom.

Four types of software sneaks are searched by analysts: the undesired inhibit of an output, wrong timing, a program message poorly describing the actual condition, and presence of an undesired output. At the discovery of sneaks, analysts conduct investigative analyses to verify that the code under consideration does indeed produce the sneaks. At the end, the impact of sneaks is assessed and appropriate corrective measures are recommended. The method is described in detail in Refs. [34–36].

15.4 Medical Device Safety

A medical device must not only be reliable but safe also. A safe medical device may simply be described as an item that does not cause too much risk to humans, equipment, or property. This section presents five important aspects of medical device safety.

15.4.1 Facts and Figures

Some of the facts and figures directly or indirectly related to medical device safety are as follows [11, 38]:

- In 1969, the U.S. Department of Health, Education, and Welfare special committee reported that over a period of 10 years, there were approximately 10,000 medical device related injuries and 731 caused deaths [39, 40].
- A sample of 15,000 products used in hospitals was examined by Emergency Care Research Institute (ECRI) and then it concluded that 4 to 6% of these items were dangerous to warrant immediate corrective action [11, 38].
- A patient was seriously injured by over-infusion because an attending health care professional wrongly read the number "7" as "1" [41].
- A five-year old child patient was crushed to death beneath the pedestal-style electric bed [42].
- A patient died due to radiation overdose involving a Therac radiation therapy device [43].

15.4.2 Medical Device Safety vs. Reliability

Although safety and reliability are quite distinct concepts, at times they can have conflicting concerns [44]. Thus, their clear understanding is absolutely essential. Nonetheless, a medical device is still considered quite safe even if it fails often without mishaps. On the other hand, if a medical device operates normally at all times but puts people at risk, under such conditions the device is considered reliable but unsafe. Two examples of safe-unreliable and reliable-unsafe medical devices, respectively, are given below [44].

- A pacemaker does not always pace at the specified programmed rate for a small number of times and is not a particular safety concern. Under such circumstances, the pacemaker is safe but unreliable (i.e., safe-unreliable).
- A pacemaker paces, say at 114 beats/min under any condition, thus is considered very reliable. However, if the patient involved

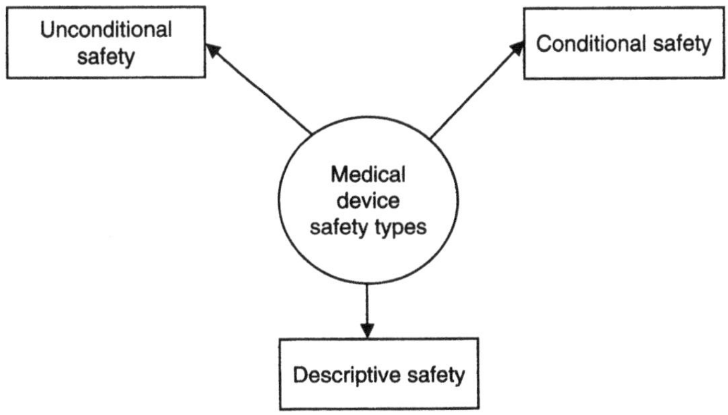

FIGURE 15.2
Medical device safety categories.

is in cardiac failure, a high pacing rate, from the medical point of view, is considered inappropriate. Under such circumstances, the pacemaker is reliable but unsafe(i.e., reliable-unsafe).

15.4.3 Types of Medical Device Safety

Medical device safety may be divided into three distinct categories as shown in Figure 15.2 [45]:

Unconditional safety is the most effective and is preferred over all other possibilities. It calls for the elimination of all possible risks through an effective design process. Conditional safety is used in circumstances where unconditional safety cannot be realized. For example, in the case of laser or x-ray devices, as dangerous radiation emissions cannot be avoided, the risk is reduced through measures such as restricting access to therapy rooms, or installing a locking power switch that allows the activation of device by authorized individuals only. Additional indirect safety measures such as x-ray folding screens, protective laser glasses, and protective (lead) x-ray skirts can also be employed.

Descriptive safety is used in circumstances when it is impossible or inappropriate to provide adequate safety through the above two means (i.e., conditional or unconditional). Descriptive safety in regard to areas such as maintenance, operation, mounting, transport, connection, and replacement may simply involve displaying signs such as:

- "Shake before use"
- "Handle with care"
- "Not for explosive zones"
- "This side up"

15.4.4 Patient Injury and Medical Device Accident Causes

Over the years professionals working in the medical field have identified many fundamental causes of patient injuries. Some of these are overdose, skin lesions (i.e., burns), electrocution, suffocation/barotraumas, fire, embolism, and crushing [46]. Similarly, they (i.e., professionals) have grouped the causes for medical device accidents under the following seven categories [46]:

- Design defect
- Manufacturing defect
- Faulty repair, preventive maintenance, or calibration
- Sabotage or malicious intent
- Abnormal or idiosyncratic patient response
- Patient/operator error
- Random component failure

15.4.5 Medical Device Safety Requirements

In order to improve the safety of medical devices, over the years government and other bodies have imposed various types of safety requirements on medical devices. These requirements may be grouped under the following three areas [45]:

- **Safe Function.** This includes requirements such as accuracy of measurements, warning against or prevention of dangerous outputs, and reliability.
- **Sufficient information.** This includes elements such as effective labelling, accompanying documentation, and instructions for use.
- **Safe design.** This includes requirements such as protection against electrical shock, protection against radiation hazards, care for environmental conditions (e.g., electromagnetic interactions, humidity, and temperature), excessive heating prevention (this requires a careful attention to items such as temperature control, cooling, and effective design), and mechanical hazard prevention (this requires attention to factors such as safe distances, device stability, and breaking strength).

15.5 Problems

1. Define the term "robot safety."
2. List at least seven safety problems unique to robots.
3. Discuss four types of robot accidents.

4. Discuss major causes for the occurrence of various types of robot hazards.
5. Discuss the following robot safeguard methods:
   - Flashing lights
   - Intelligent systems
   - Infrared light arrays
6. Define the term "software safety."
7. List at least seven software hazard causing ways.
8. Describe the following software hazard analysis methods:
   - Code walk-throughs
   - Proof of correctness
   - Software sneak circuit analysis
9. Discuss the following two types of medical device safety:
   - Unconditional safety
   - Conditional safety
10. List at least seven causes for the occurrence of medical device accidents.

References

1. Dhillon, B.S., *Robot reliability and safety,* Springer-Verlag, New York, 1991.
2. An Interpretation of the technical guidance on safety standards in the use, etc., of industrial robots, Japanese Industrial Safety and Health Association, Tokyo, 1985.
3. Bonney, M.C. and Yong, Y.F., Ed., *Robot safety,* Springer-Verlag, New York, 1985.
4. American National Standard for Industrial Robots and Robot Systems: Safety Requirements, ANSI/RIA R15.06-1986, American National Standards Institute (ANSI), New York, 1986.
5. Dhillon, B.S., Fashandi, A.R.M., and Liu, K.L., Robot systems reliability and safety: a review, *Journal of Quality in Maintenance Engineering,* Vol. 8, No. 3, 2002, pp. 170–212.
6. Keene, S.J., Software reliability concepts, *Annual Reliability and Maintainability Symposium Tutorial Notes,* 1992, pp. 1–21.
7. Cha, S.S., Management aspect of software safety, *Proceedings of the Eight Annual Conference on Computer Assurance,* 1993, pp. 35–40.
8. Russell, C., Human errors: avoidable mistakes kill 100,000 patients a Year, *The Washington Post,* February 18, 1992, pp. WH7.
9. Dhillon, B.S., *Medical Device Reliability and Associated Areas,* CRC Press, Boca Raton, FL, 2000.
10. Graham, J.H., *Overview of Robot Safety, Reliability, and Human Factors Issues in Safety, Reliability, and Human Factors in Robotic Systems,* edited by J.H. Graham, Van Nostrand Reinhold, 1991, pp. 1–10.

11. Dhillon, B.S., *Engineering Safety: Fundamentals, Techniques, and Applications*, World Scientific Publishing, River Edge, NJ, 2003.
12. Nagamachi, M., Ten fatal accidents due to robots in Japan, in *Ergonomics of Hybrid Automated Systems*, edited by W. Karwowski, et al. Elsevier, Amsterdam, 1988, pp. 391–396.
13. Jiang, B.C. and Gainer, C.A., A cause and effect analysis of robot accidents, *Journal of Occupational Accidents*, Vol. 9, 1987, pp. 27–45.
14. Sanderson, L.M., Collins, J.N., and McGlothlin, J.D., Robot-related fatality involving an U.S. manufacturing plant employee: Case Report and Recommendations, *Journal of Occupational Accidents*, Vol. 8, 1986, pp. 13–23.
15. Dhillon, B.S., *Robot Reliability and Safety*, Springer-Verlag, New York, 1991.
16. Backtrom, T. and Doos, M., A comparative study of occupational accidents in industries with advanced manufacturing technology, *International Journal of Human Factors in Manufacturing*, Vol. 5, 1995, pp. 267–282.
17. Clark, D.R. and Lehto, M.R., Reliability, Maintenance, and Safety of Robots, in *Handbook of Industrial Robotics*, edited by S.Y. Nof, John Wiley & Sons, New York, 1999, pp. 717–753.
18. Van Deest, Robot safety: a potential crisis, *Professional Safety*, January 1984, pp. 40–42.
19. Industrial Robots and Robot System Safety, Chapter 4, in *OSHA Technical Manual, Occupational Safety and Health Administration* (OSHA), Department of Labor, Washington, D.C., 2001.
20. Ziskovsky, J.P., Working Safely with industrial robots, *Plant Engineering*, May 1984, pp. 40–42.
21. Addison, J.H., Robotic Safety Systems and Methods: Savannah River Site, Report No. DPST-84-907 (DE-35-008261), December 1984, issued by E.I. du Pont de Nemours & Co., Savannah River Laboratory, Aiken, South Carolina.
22. Schneider, P. and Hines, M.L.A., Classification of medical software, *Proceedings of the IEEE Symposium on Applied Computing*, 1990, pp. 20–27.
23. Joyce, E., Software bugs: a matter of life and liability, *Datamation*, Vol. 33, No. 10, 1987, pp. 88–92.
24. Gowen, L.D. and Yap, M.Y., Traditional software development's effects on safety, *Proceedings of the 6th Annual IEEE Symposium on Computer-Based Medical Systems*, 1993, pp. 58–63.
25. Mendis, K.S., Software safety and its relation to software quality assurance, in *Handbook of software quality assurance*, edited by G.G. Schulmeyer and J.E., McManus, Prentice Hall, Upper Saddle River, NJ, 1999, pp. 669–679.
26. Anonymous, Blow Balloons, *Aviation Week Space Technology*, September 20, 1971, pp. 17.
27. Thayer, R.H., Software Engineering Project Management, in *Software Engineering*, edited by M. Dorfman and R.H. Thayer, IEEE Computer Society Press, Los Alamitos, CA, 1997, pp. 358–371.
28. Leveson, N.G., Software safety: why, what, and how, *Computing Surveys*, Vol. 18, No. 2, 1986, pp. 125–163.
29. Ericson, C.A., Software and system safety, *Proceedings of the 5th International System Safety Conference*, 1981, pp. III B.1–IIIB.11.
30. Roland, H.E. and Moriarty, B., *System Safety Engineering and Management*, John Wiley & Sons, New York, 1983.
31. Leveson, N.G., Software safety in computer-controlled systems, *IEEE Computer*, February 1984, pp. 48–55.

32. Friedman, M.A. and Voas, J.M., *Software Assessment*, John Wiley & Sons, New York, 1995.
33. Leveson, N.G., *Safeware*, Addison-Wesley, Reading, MA, 1995.
34. Ippolito, L.M. and Wallace, D.R., A study on hazard analysis in high integrity software standards and guidelines, Report No. NISTIR 5589, National Institute of Standards and Technology, U.S. Department of Commerce, Washington, D.C., January 1995.
35. Hammer, W. and Price, D., *Occupational Safety Management and Engineering*, Prentice Hall, Upper Saddle River, NJ, 2001.
36. Hansen, M.D., Survey of available software-safety analysis techniques, *Proceedings of the Annual Reliability and Maintainability Symposium*, 1989, pp. 46–49.
37. Sheriffs, Y.S., Software safety analysis: the characteristics of efficient technical walkthroughs, *Microelectronics and Reliability*, Vol. 32, No. 3, 1992, pp. 407–414.
38. Dhillon, B.S., *Medical Device Reliability and Associated Areas*, CRC Press, Boca Raton, FL, 2000.
39. Banta, H.D., The regulation of medical devices, *Preventive Medicine*, Vol. 19, 1990, pp. 693–699.
40. Medical Devices, Hearings before the Subcommittee on Public Health and Environment, U.S. Congress Interstate and Foreign Commerce, Serial No. 93–61, U.S. Government Printing Office, Washington, D.C., 1973.
41. Sawyer, D., *Do it by design: introduction to human factors in medical devices*, Center for Devices and Radiological Health (CDRH), Food and Drug Administration (FDA), Washington, D.C., 1996.
42. *Electric Beds Can Kill Children*, Medical Device Safety Report, prepared by the Emergency Care Research Institute (ECRI), Plymouth Meeting, Pennsylvania, 2001.
43. Casey, S., Set Phasers on stun and other true tales of design technology and human error, *Aegean*, Santa Barbara, CA, 1993.
44. Fries, R.C., *Reliable Design of Medical Devices*, Marcel Dekker, New York, 1997.
45. Leitgeb, N., *Safety in Electro-Medical Technology*, Interpharm Press, Buffalo Grove, IL, 1996.
46. Brueley, M.E., Ergonomics and error: who is responsible?, *Proceedings of the First Symposium on Human Factors in Medical Devices*, 1989, pp. 6–10.

Index

Q

R

S

T

U

V

W

For Product Safety Concerns and Information please contact our EU representative GPSR@taylorandfrancis.com
Taylor & Francis Verlag GmbH, Kaufingerstraße 24, 80331 München, Germany

www.ingramcontent.com/pod-product-compliance
Ingram Content Group UK Ltd.
Pitfield, Milton Keynes, MK11 3LW, UK
UKHW011005180926
475701UK00004B/572

* 9 7 8 0 3 6 7 3 9 3 6 0 1 *